ISBN 978-1-943492-44-2

Book and cover design by **designpanache**
Cover photograph by Mick J Prodger

ELM GROVE PUBLISHING
San Antonio, Texas, USA
www.elmgrovepublishing.com

This book is dedicated to the late George Zielinski, A.G.P., one of the greatest men I've ever known.

Praise for Fall of the Japanese Empire

"The publication, in 2008, of Ron Werneth's Beyond Pearl Harbor, The Untold Stories of Japan's Naval Airmen *constituted a true milestone in the history of aviation in the Pacific War. Very rarely had the Japanese perspective emerged so masterfully through numerous personal interviews carefully placed in proper context with extensive background research. Now, ten years later, he has repeated his earlier accomplishment with another outstanding volume, with Allied as well as Japanese recollections, that takes the story to the end of the war. Time has silenced the voices of nearly all of the remaining World War II aviators. Those of us who seek to learn more about that terrible conflict owe Ron Werneth a great debt for his lifelong devotion to getting to know some of these remarkable men and ably preserving their accounts."*

– John B. Lundstrom, author of *The First Team*

"Ron Werneth's dedication to preserving the first-person accounts of Pacific War veterans, particularly Japanese aviators, is apparent on every page of this remarkable compilation. His interviews with more than 60 Japanese veterans contribute to a unique perspective that few Western historians could ever hope to gain."

– Bruce Gamble, author and historian

"As a forensic historian, Ron Werneth sets up the battle situation, then lets the veteran tell their stories in their own words. It was as if the vet was talking directly to me! Fall of the Japanese Empire *is not a reader unfriendly history book; rather, it is the personal and interesting accounts of men who are now no longer here. These were their last words, spoken with wisdom and emotion."*

– Henry Sakaida

"A second superb book from a superb researcher. Werneth had already demonstrated his passion for getting unique stories—moving to Japan and living there for several years in order to interview as many of Japan's surviving WWII aviators as he could. This book does the same, but adds U.S. aviators into the mix for balance and perspective. A must-have for any Pacific War aviation enthusiast!"

– Jon Parshall, author and historian

Explanatory Note about the Text

All times listed in the veterans' accounts are local times. The author chose to use some Japanese military terms and words to make the text as authentic as possible. Please see the glossary (page 262) for translations of Japanese terms used. All of the Japanese participants in this work are presented with their given names first, then their family names. The author frequently uses "san" after a Japanese family name as this is the polite way it is done in their native language. Lastly, the Japanese language word for Navy Air Group, *Kōkūtai* or its abbreviated form, *Ku*, is used throughout the text, because this is the term the veterans themselves used in both oral interviews or written text sources. All of these airmens' stories are told in their own words, as recorded by the author, or as they were translated from the original interviews and in some cases transcribed from diaries and other personal documents. Throughout the text, the author's narrative appears in *italic.*

Table of Contents

Acknowledgments

A BOOK PROJECT SUCH as this wouldn't be possible without the help of so many wonderful people who supported me throughout the fifteen-year journey that took me around the world to Japan (for a three-year stay) along with the subsequent research over the past twelve years. This "journey" resulted in two books, this one, and my first work, *Beyond Pearl Harbor: The Untold Stories of Japan's Naval Airmen.*

In the United States, the following individuals graciously supported me, in some cases, for many years, to help make this book a reality. Historians Bruce Gamble, James Lansdale, Michael Lemish, John B. Lundstrom, Jonathan Parshall, Henry Sakaida, James Sawruk, Anthony Tully and J. Michael Wenger. All of these gentlemen played significant roles from helping me research veterans' accounts to proofreading passages in the text to make sure all of the information presented could be as accurate as possible. While based back in Chicago, I received enormous assistance in the proper translation of several vital Japanese veterans' diaries and related resources from my good friends Koji Ikenaga and Masami Takahashi. I am thankful for the support of several professional friends who helped me along the way including: Daniel Burke, Duncan Cameron, John Conway, John Egger, John Lyon, Dave Marks, Mark Stone, Ed Tracy and John Zukowsky. I would like to thank historian Stephen L. Moore who kindly provided wartime era documents and photographs related to Torpedo Squadron Ten.

For the aircraft profiles and related artwork, I appreciate the hard work of Michael Claringbould (in Australia) and the late George Zielinski A.G.P. Their attention to detail was second to none. I would like to thank my friends and aviation aircraft masters Derek Brown, Bobby McElroy Jr. and Ryan Toews (Canada), for their assistance with the aircraft profiles. I appreciate that aviation artist Roy Grinnell and his wife Irene kindly allowed me to use his painting "Mission Accomplished" which depicts the Yamamoto Mission for this book. Last, but not least, many thanks to UK-based aviation artist Rod Kirkby for allowing me to use his painting "Final Combat" that illustrates the last engagement on 15 August 1945 between British Royal Navy fighters and their Japanese Imperial Navy counterparts.

In Japan, I feel so fortunate to have had the assistance of historians Minoru Kamada, Takehikō Shibata and Ikuo Shimizu. These gentlemen played vital roles in the proper research and translation of numerous Japanese-language resources including wartime combat reports, veterans' diaries/correspondence and other helpful publications. Without their help, this book could have never been properly researched from the Japanese-side. Another essential person to this work was my longtime translator Keigo "Hammer" Nakahama. Hammer was responsible for translating my personal interviews with the majority of Japanese veterans. I am indebted to him because I would have never been able to meet these men without his help (for so many years). I appreciate the kindness of aviation photo specialist S. Bon Group and allowing me to use numerous rare wartime images from their incredible photo archives. I also would like to thank aviation historian Hideki Yamauchi and his friend T. Kamino who allowed me to publish several of his great Japanese Imperial Navy ship images in this book.

Without the last "essential" part of this book – the support and willingness to participate of both the Japanese and American World War II veterans, this unique collaboration would have never happened at all. Thank you to all of these men and their families who made this possible.

From the United States: Ed Adamski, David Braden, Doug Canning, George Center, Charles Downey, John "Dugan" Doyle, Ed Fox, Mert Hansen and the Hansen family, Fred Olivi, Emily Manchester Townes (sister of VB-80 veteran John Manchester), George Walsh and Grant C.Young.

From Japan: Saburō Abe, Fujio Hayashi, Kunio Iwashita, Atsumi Kozono and the Kozono family, Shigeō Nakagawa, Shinichi Ōguro, Kozo Ōhkawa, Yoshio Ōshitani, Minoru Toyonaga, Ryōichi Yamada, Mitsuyasu Yamakawa and the Yamakawa family, Kenji Yanagiya, Minoru Yoshino and the Yoshino family and Morimasa Yunokawa.

I am grateful and indebted to my longtime friend Mick Prodger and his firm Elm Grove Publishing for "understanding" the work I wanted to publish and making it a reality.

Last, but not least, I could not have completed this book without the support and love from my beautiful wife Kaori and our daughter Maya.

RW

One of the most difficult things for a writer, especially one who is not fluent in the language, is proper research. Two Japanese historians, Ikuo "Ike" Shimizu (left of the author) and Minoru Kamada, provided essential research from numerous resources, as well as a clear understanding of official Japanese archival documents and other resources such as veterans' diaries, for the background of this book. This picture was taken during a visit to The National Institute for Defense Studies in Tokyo, Japan. (Ron Werneth)

Keigo "Hammer" Nakahama has been my friend and translator in Japan since the early days of my research in Japan. He was responsible for translating the majority of the Japanese veteran interviews and numerous primary source materials. Here he (left) is pictured with Lieutenant (junior grade) Saburō Abe (right) after an interview with the author. (Ron Werneth)

Introduction

Over seventy years ago World War II ended with the staggering loss of over 60 million people, which in 1940, was roughly 3% of the world's population.[1] To put that in perspective, imagine two entire American states, California and Florida, being obliterated in front of your eyes. In one instant, everyone living in these two states would be gone.

With the passing of over seven decades of time, the facts and stories of World War II may be misunderstood and totally reinterpreted by later generations of people. This is especially true if future students of history don't have an understanding of "both" sides of the war. Case in point, look at the current media stories about the American use of the atomic bombs in WW II against the nation of Japan. It is safe to say if you talked to veterans of this bloody conflict, they would feel totally differently about the use of this terrible weapon that ultimately ended the fighting. I will never forget the frank conversation I had with WW II Japanese Navy aviator and Pearl Harbor veteran Takeshi Maeda. In short, he said that these nuclear weapons saved his life because Maeda-san would have been expected to make a suicide attack with his aircraft on the impending Allied invasion forces of the mainland of Japan. At that late stage of the war, the military forces of Japan were conserving most of their aircraft so they could do "mass" suicide air attacks during an invasion. I believe the majority of the sixty plus Japanese WW II veterans I interviewed felt that Japan would have never surrendered without the dropping of these lethal new weapons of war. Even after these horrific bombs were used and the Emperor did a formal radio announcement of the surrender of his nation, many of Japan's veterans couldn't even fathom this and thought it must be a mistake of some kind. Because of this, many Japanese military men vowed to continue the fight such as the elite 302 *Kōkūtai* (Japanese Navy Air Group) led by career naval officer, Captain Yasuna Kozono. As you read through this work, you can hear some of these firsthand accounts from the people who lived through this unstable time period at the end of the war. I hope you will come to your own conclusions.

Over twenty years ago, I started working with the World War II veterans in the United States. This research ultimately led to me moving to Japan literally out of the frustration that I could not find many good Japanese veteran stories (in English) about what really happened in the war. By this time, I was fortunate to have met and interviewed many American WW II veterans but getting the opportunity to do the same thing in Japan required many years of work beforehand. It also was essential to have the assistance of a handful of wonderful people that *understood* why I wanted to do this – mainly, in the United States, Japanese American historian Henry Sakaida, and in Japan, the late Jirō Yoshida, a former WW II Japanese Navy pilot. They both were instrumental for getting this project started in the late 1990s, then I took the plunge and moved to Japan.

All together, I spent over three years living and working in Japan. When I wasn't working as an English teacher (to pay the bills), I was often busy interviewing veterans all across Japan. I felt so fortunate to gain the trust, and oftentimes, close friendship of these former naval aviators, aircraft mechanics and ground personnel that I got to know so well. Because of this, I was able to attend several important reunions such as the ones held for the 302 *Kōkūtai, Jinrai Butai* (Divine Thunder Group) and the last "official" meeting of the Zero Fighter Pilots Association. Furthermore, several of these Japanese veterans became such close buddies that they attended my wedding on 5 October 2003, in Tokyo.

After I published my first book, *Beyond Pearl Harbor: The Untold Stories of Japan's Naval Airmen*, which focused mainly on the early part of WWII (mostly from the Japanese perspective) I realized that another work, as told from the eyes of both perspectives, the American and Japanese veterans, needed to be done. During my over two decades of work with these amazing men, I had a lot of stories to share but my time with them was running short due to their advancing age so I initially focused all of my efforts on interviewing as many of them as possible before it became too late. Therefore, this second book took longer to publish than I wanted it to. However, it was essential to do this work *right* with proper translations which were supported by official documents in the respective national archives including: Japan and the United States along with its wartime allies, Australia, New Zealand and the United Kingdom.

This book focuses on the time period of late 1942 until the end of the war. Japanese Navy Admiral Isoruko Yamamoto, was very prophetic in his belief what would happen if a war occurred with the United States and its allies. He said, "If we were ordered to do it, then I can guarantee to put up a tough fight for the first six months, but I have absolutely no confidence as to what would happen if it went on for two or three years."[2] Admiral Yamamoto felt that attacking Pearl Harbor and utilizing his naval airpower and aircraft carriers to their fullest potential would be the *only way* Japan could have any chance in a war of this magnitude against the United States, an industrial large nation. Just as he predicted, after the successful Pearl Harbor attack, a string of subsequent Japanese victories followed such as in Burma (now called Myanmar), the Dutch East Indies and the Philippines. However, during the summer of 1942, things started to change for the Japanese with their stinging losses during the Battle of Midway followed by the Allied invasion of Guadalcanal - the first of their Pacific island hopping campaign.

1943 wasn't much better for the military forces of Japan. In February of that year, Japanese forces evacuated Guadalcanal and never really gained any offensive momentum after that. Then, just two months later, on 18 April, Admiral Yamamoto himself became a casualty of war when the aircraft he was flying in was shot down. This is really where our story begins with Japan on the defensive, and what it was like being a participant during the final two years of World War II in the Pacific. I *purposely* focused this work on the firsthand stories of the people who lived through this pivotal time in history; not just one specific battle or campaign – plenty of other books can do that. It is my sincere hope that these vignettes of amazing WWII veterans promote a better understanding of what war is really like and why future armed conflicts should be avoided at all cost.

[1] www.census.gov/

[2] Hiroyuki Agawa, *The Reluctant Admiral,* Yamamoto and the Imperial Navy

Chapter 1

"Flying for the Air Force was my job and I loved it."

Lieutenant Colonel Douglas Canning (USAAF)

While I was researching the famous Yamamoto Mission (when the well-known Japanese admiral was shot down) I sadly learned that there was only one living American fighter pilot who participated in the engagement, Douglas Canning. I heard that he flew combat missions using the P-39/P-400 fighter over Guadalcanal and thought it would be important to include his story. With the help of Florida-based historian Jim Lansdale, I was able to interview him. Canning was very humble but saw a lot of action during three wars; I feel very fortunate to have talked with this distinguished aviator.

Early Years

I was born in Wayne, Nebraska on 4 July 1919. I did CPT (Civilian Pilot Training) before joining the United States Army Air Forces. I graduated from Kelly Field on 26 September 1941 while in class 41-G. From there, I went to Hamilton Field to the 70th Pursuit Squadron and we (the 70th) were on board a ship that was supposed to set sail on 7 December 1941. Our squadron was going to join the 35th Fighter Group in the Philippines. After the Pearl Harbor attack, we went back to Hamilton Field, where we got six hours flying time in a P-40 fighter. In early January 1942, we sailed for twenty-three days to our destination, which we later learned was Suva, Fiji. The aircraft there were P-39Ds. These fighters were assembled at Latoka and then we went to Nandi Airport. At Nandi, we trained in fighter tactics until 15 September with Captain John Mitchell. He ended up picking fourteen of us to go with him to 67th Fighter Squadron at Tontouta, New Caledonia.

On 15 May 1942, both the 67th and 70th pursuit squadrons were redesignated as fighter squadrons.[1]

Flying the P-400 over Guadalcanal

On 1 October, seven of us went to Henderson Field (on Guadalcanal Island) to fly P-400s, a lousy version of the P-39, which had been rejected by the British. At Henderson, I flew my first tour (in P-400s) followed by my second and third tours in P-38s. Both the P-400 and P-39s were a lot of fun to fly, but the *Zero* fighters were much faster and could outturn them. In order to have a chance against the *Zeros*, you really had to be above and behind them then swoop down to make your attack. This is because you couldn't turn with them and they could do two turns (versus one in our aircraft) then they would be right on your tail. Nevertheless, I thought the P-39 was a good aircraft to fly, but not suitable for air-to-air combat. I think, however, our squadron probably shot down five or six *Zero* fighters while flying the aircraft. The P-400 was equipped with a high-pressure British oxygen system, which couldn't be serviced in the Pacific Theatre. Therefore, you could only fly as high as normal breathing would allow, so typically, 10,000 feet was your maximum altitude.

P-39s had a low pressure system and we lightened one as a test-aircraft and it was able to go up to 28,000 feet. The P-400 flew the same as a P-39, but was a bit drafty. It did have a smaller engine so was slower but had the same number of guns, two .50s and four .30 calibers. The aircraft also could carry bombs under the fuselage. In October 1942, the P-38s arrived and airmen from the 67th and 70th fighter squadrons were hand-picked to fly the new fighters with the newly formed 339th Fighter Squadron.

Members of the 67th Fighter Squadron, the "Fighting Cocks" who are standing in front of one of the unit's P-400 fighters, named "Hells Bell" (a pun on the aircraft manufacturer Bell). Left to right, near the cockpit area: F. Williams (pith helmet) and A. Fitzgerald (shirtless with a cap), both were armorers. Back row: James Jarman (cap), Jerome Sawyer (t-shirt) and Ray Campbell. Second row: Albert Farquharson, Robert Ferguson and Leland Ramp. First row: K.S. Fjelstad, George Dewey, Besby Holmes, James Morton, Delton Goerke, Douglas Canning, Robert Kerstetter, M.C. Haedtler and Danforth Miller. All the unit members in this picture are lieutenants. (347th Fighter Group Association via Jim Lansdale)

I did no air-to-air combat while flying the P-400. While I was on Guadalcanal, I flew strafing and dive-bombing missions with the aircraft. When I first arrived on the island, we owned just about a half-mile deep by one and a half-mile long area, which included Henderson Field. So, as soon as we took off from Henderson Field and got a little altitude, we were already over the second line of defense. Therefore, we could almost immediately start shooting after we were in the air. The frontline Marines put cerise panels in front of their positions.[2] My guess is they were three feet by eight feet in size. Beyond the panel it was a free firing zone for bullets, bombs, cannon fire and fire bombs (the forerunner of napalm).

We got bombed almost every day and night by Japanese bombers and *Zero* fighters. So, our men had foxholes everywhere and finally on my third tour, I had a foxhole right in my tent! We thought the Japanese had one five-inch artillery piece out there but when we got to it, we actually found two. Our men called them "Pistol Pete." Once you heard them fire, you had ten seconds to get into a foxhole for cover. Towards the end of my first tour, we slept in a cave for protection. Initially, there were a lot more Japanese troops on Guadalcanal, but we did such a good job, the enemy eventually evacuated the island in February 1943.

My second tour on Guadalcanal was as a member of the 339th Fighter Squadron (P-38s) where I flew half a tour on Fighter 1 and the second half on Fighter 2. My third and final tour was out of Fighter 2 (339th, P-38s).

Combat Missions with the P-38 fighter

The P-38 fighter was my favorite aircraft to fly (and I even flew jets later on in my flying career). I got a model of one in my bedroom. Initially, we flew "F" model P-38s, but later got the "J" variant with the aileron boost. We had a control wheel on the P-38 and it took a lot of muscle to fly the fighter. However, with the addition of the aileron boost in the "J" version, it made the fighter a lot easier to fly. We flew a lot of escort missions with our bombers when they went to attack Bougainville Island. It was maybe 400 miles each way. Normally, we used 165-gallon belly tanks on each wing. However, for the Yamamoto Mission we carried one 165-gallon and one 310-gallon tank.[3] At normal cruising speeds, the engines would consume seventy-five gallons per hour and they used 100 octane fuel.

I thought using the P-38 to fight enemy fighters was similar to the P-39. Once I got on the tail of a *Zero* fighter, he could make two turns and was shooting at me! On one occasion this occurred, I dove away, but he got a bullet in my wing. I thought our fighters at that time including the P-38, P-39 and P-40s, couldn't turn with the *Zero* at all. The enemy pilots were very skillful, but in stupidity had their radios, self-sealing fuel tanks and armor plate removed, which made them climb and turn better.[4] Because of these factors, we didn't climb and turn with them. We dove in one pass and pull up and then, if lucky, we made a second diving attack. The main problem is that when you hit them (enemy aircraft), they blew up and you were in danger of being hit by their debris.

I remember a mission (which occurred on my third tour, around 1943) when we burned an enemy ship. One afternoon, Lieutenant Colonel Henry Viccellio sent seven of us (P-38s) to drop belly tanks on a transport in a harbor two hundred miles away. Captain Tom Lanphier led the first flight (three fighters) and I led the second flight (four aircraft). We had been dropping belly tanks during our first tour on the Japanese troops because they blew up at the top of the jungle, but the fire would penetrate the jungle. (On this mission) my belly tank hit the ship's bridge and gasoline was all over. After this, my wingman fired at the bridge and the ship started to burn. Around midnight, a PBY Catalina saw it explode violently so evidently the ship carried munitions. We got two messages from Admiral Halsey by name, one congratulating us on our big time hotfoot and the other one stating that our trick was neat and hard to beat. He also gave his congratulations and a case of whiskey! After the Yamamoto Mission, he sent us two cases. During the war, many people thought and we still do that this is how the idea of napalm came about.

Yamamoto Mission

On Sunday, 18 April 1943, Canning participated in one of most important aerial engagements of World War II, the interception and shoot-down of Japanese Navy hero and the leader of the Combined Fleet, Admiral Isoroku Yamamoto. This important mission has been exhaustively researched and documented in several books (please see the bibliography if you like to read more about this engagement). I wanted to "focus" on veteran Canning's experiences for this account.

In short, U.S. Navy code-breakers learned that Admiral Yamamoto would be doing an inspection and try to "rally his men" at several forward bases in conjunction with the recently launched I-Go Operation, which was a "move" to neutralize some of the allied air bases/offensive moves in the Solomon Island chain (such as Henderson Field on Guadalcanal Island). Yamamoto, who was known to be always punctual, was planning to visit (on 18 April) the following bases: first, Ballale Island, then Shortland Island and lastly Buin (on Bougainville Island). During some of this trip, the admiral would be traveling via a subcasher (island to island), the leader of the mission, Major John Mitchell felt that his best opportunity to "bag his prey" would be in the air as he wasn't as familiar with ship identification and that Yamamoto may have a better chance of survival if he was on board a ship. Since none of the current American naval fighters had the range to complete this mission, the only airmen and aircraft suitable for this task were the 347th and 18th Fighter Groups with their twin-boom P-38 Lightning fighters. As Canning mentioned earlier, instead of P-38s being fitted with two auxiliary 165-gallon belly tanks (to supply the extra fuel), they had to utilize one 165-gallon and an additional 310-tank to complete his long-distance interception.

Legendary Admiral Isoroku Yamamoto, leader of the Imperial Japanese Navy Combined Fleet. He was a career Navy man but knew that the future of military conflicts would be dictated by the power of aircraft, not the battleship. (NA)

The eighteen airmen from both fighter groups were to be chosen by respective commanders Lieutenant Colonel. Henry Viccellio (347th) and Lieutenant Colonel Aaron Tyler, but Major Mitchell and Major Louis Kittel (the acting commanding officer of the 12th Fighter Squadron) made up the roster of the pilots who would participate in the historic mission. Mitchell planned to use his eighteen P-38s in this fashion, fourteen of them would stay at a high-altitude (18,000-20,000 feet) while the remaining four fighters would go after Admiral Yamamoto and his escorts. Long-time Yamamoto Mission researcher Jim Lansdale determined that the outbound trip from Fighter Two (on Guadalcanal Island) was around 450 miles (+/-20 miles) with the return leg being around 400 miles. So with the total distance flown (on the mission) being a little less than 900 miles, this was the longest distance ever flown on a fighter intercept mission during the war. Outbound, the P-38 fighters had to fly an unusual plotting (which was done to avoid enemy coastwatchers). Major Mitchell had to do meticulous planning in order to calculate the speed/distance of every part of this mission, including the anticipated travel time of Yamamoto's aircraft. Lastly, in order to help "guide" his group for this long-distance flight, he also requested that his personal P-38 be specially fitted with a U.S. Navy ship's compass.

At 0710, on Easter Sunday, eighteen P-38 Lightning fighters, led by Major John Mitchell roared to life on Fighter Two and began the longest aerial interception mission of World War II. Unfortunately, two P-38s suffered mishaps soon after firing up with their twin V-1710 Allison engines. While taxiing his aircraft, First Lieutenant Jim McLanahan had a blown tire after hitting a jagged piece of Marston matting and was stranded on the runway. Next, after getting airborne, First Lieutenant Joe Moore couldn't get his 310-gallon belly tank to feed so he too, had to reluctantly abort from this mission. Both McLanahan and Moore were assigned to a flight of four P-38s that were the "attack" group so Mitchell signaled First Lieutenant Besby Holmes and First Lieutenant Raymond Hine to join Captain Thomas Lanphier, Jr. and First Lieutenant Rex Barber to take their spots in the "hunter" section.

Canning doesn't mention it in this version of his mission, but he was the pilot who first spotted the two Mitsubishi G4M1 (Allied code name "Betty") bombers (utilized as transports), from the 705 Kōkūtai *(or* Ku*) and six escorting Zero fighters (204* Ku*). One bomber had eleven passengers including seven aircrew, and four passengers, Admiral Yamamoto and his staff. The other* Betty *had twelve people on board including seven aircrew, and five passengers including Vice-Admiral Matome Ugaki and four staff officers. After spotting the Japanese aircraft, Canning called out, "Bogeys. Eleven o'clock high." At first, Mitchell was surprised because the initial Japanese coded-message stated that there would be just one bomber, not two, but after seeing the six enemy fighters escorting the two aircraft,* *he knew that they found their "targets."*

Upon realizing these enemy aircraft were the "correct" targets, Mitchell and eleven other pilots (including Canning) climbed to higher-altitude (15,000-18,000 feet) to keep vigil for the possibility of additional Japanese fighters coming up to engage them. At the same time. the "attack" group of four P-38s, flown by (Lanphier, Jr., Barber, Holmes and Hine) planned to go after Yamamoto's bomber and the other Japanese aircraft. However, Holmes couldn't drop his belly tanks so he turned southeast along the coast of Bougainville Island trying to shake them loose. Therefore, Hine also broke away from the attack flight and flew as wingman for Holmes. Because of this problem, the attack flight was once again down to just two airmen, Captain Thomas Lanphier, Jr. and First Lieutenant Rex Barber. What happened next would become an "unresolved controversy" for many years after the war.[5]

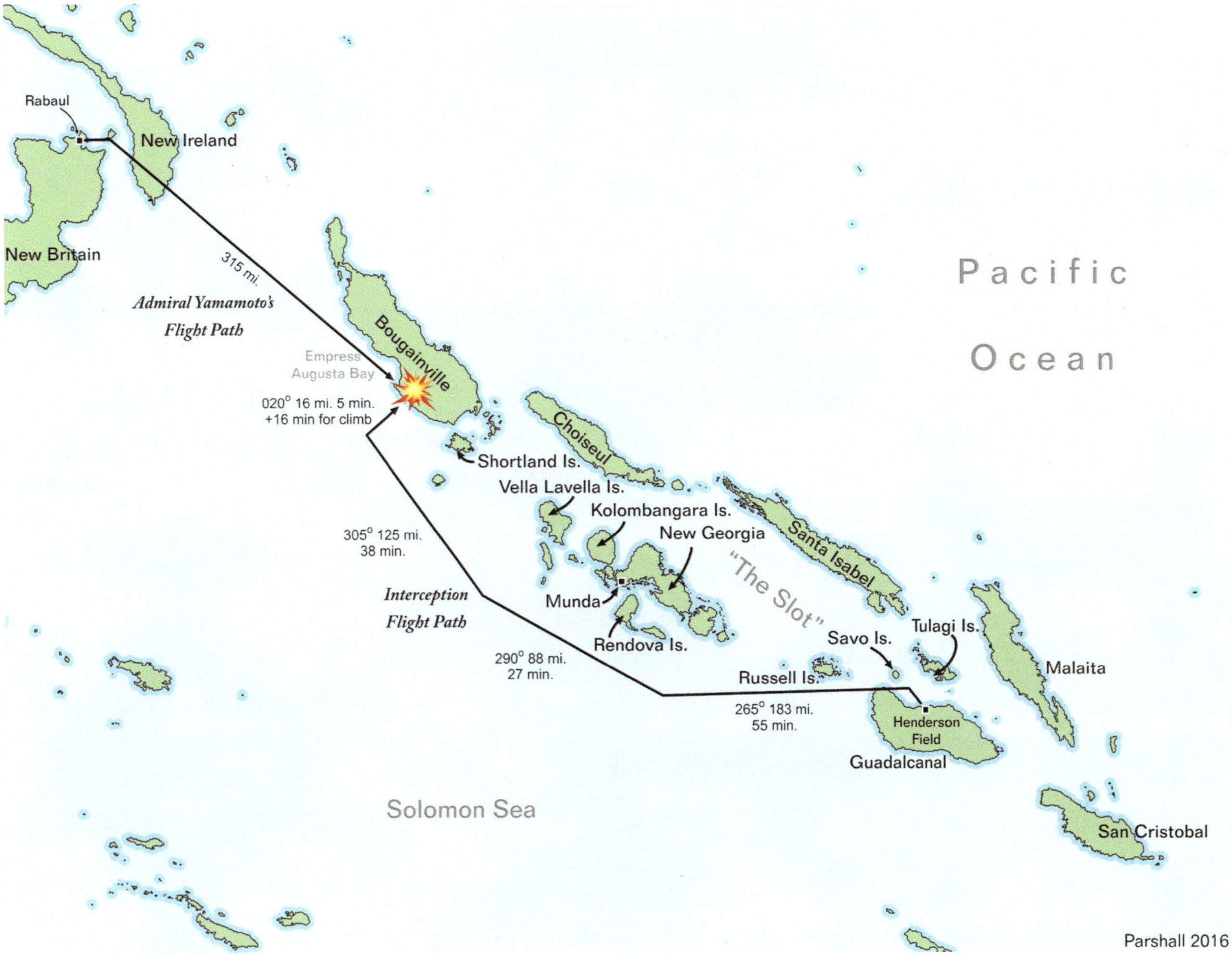

Map showing the flight courses of the group of Japanese aircraft including the one Admiral Yamamoto was flying in and the opposing sixteen P-38G Lightning fighters led by Major John Mitchell, commanding officer of the 339th Fighter Squadron. (Illustration by Jon Parshall)

*****

Typically when high-ranking officers came to Guadalcanal, we came out 100 miles or so to escort them to our base so they would have some protection. We did this duty with twenty-thirty aircraft but the Japanese Navy only assigned six *Zero* fighters to escort Admiral Yamamoto. Four of our P-38s went to attack their *Betty* bombers while twelve of us (including myself) went towards Kahili to engage any enemy aircraft that took off.

On our way towards the target area we were just fifty feet above the water. Then, we climbed and leveled off around 18,000 feet. I saw a *Zero* fighter coming around and trying to get behind us. So, I dove down and tried to get on his tail. I descended about 4,000 feet and was then flying at 14,000 feet. I had this guy bore-sighted and he was dead except for one thing, my canopy was fogged up so I couldn't see anything. Because of this, all I could do was pull up and get away. I don't think the enemy pilot even knew I was around. I went back up to 20,000 feet and looked around, but I couldn't see any other aircraft, either friendly or enemy, in the sky. I turned to go home and after about fifteen minutes or so, I caught up with First Lieutenant Besby Holmes. Since he had been flying at full throttle for quite some time, his engines were running rough and the P-38 was shaking so bad that Besby couldn't even read the instruments. Luckily, by looking through my props at his props, I could determine the correct rpm and gave him the readings.

I stayed with him because he was low on fuel and I thought he was going into the drink. If he did, I wanted to make sure of the ditching location so later I could help him get rescued. When we were about 100 miles away from Guadalcanal, there was an air base being built (in the Russell Islands), but at that time only about 2,800 feet was completed. I went down to buzz the runway so the workers would clear off, but this didn't help. So I buzzed the runway again (much lower) and this worked. Besby came in just underneath me and made a safe landing. When they measured his fuel, he had only four gallons left. Right next to that island, there was a PT Boat base there and they sent boats over with five gallon drums of fuel so he could get home.

I was the last one to land that day, but Besby returned the following day. Prior to the mission when we had our briefing, it was supposed to be top secret that the Japanese codes were broken but there were about forty people present. We were told that if anything was ever said about the codes being broken, we would be court-martialed. But, the word was out and when Captain Lanphier got on the ground, he was yelling that he got Yamamoto. I always thought Lanphier was the pilot who shot down Yamamoto until I met a Japanese Yamamoto Mission veteran, Kenji Yanagiya, at the Nimitz Museum (now named The National Museum of the Pacific War) symposium. Then, when I heard the Japanese side of the story, I changed my mind.

*****

WHILE FIRST LIEUTENANT Besby Holmes and First Lieutenant Raymond Hine were dealing with Holme's hung up belly tanks, Captain Thomas Lanphier, Jr. and First Lieutenant Rex Barber went after their prey, the two Mitsubishi G4M1 bombers, which were in descending altitude. One of the bombers, tail number 323, was transporting Admiral Isoroku Yamamoto and his staff and the second aircraft, tail number 326, carried Vice-Admiral Matome Ugaki (Chief of Staff) along with a number of other Navy officers. As the P-38s approached the flight of two bombers and six Zero *fighter escorts, the latter Type 0s released their belly tanks and prepared to engage them. At this time, the two Mitsubishi bombers, separated, with Yamamoto's aircraft heading over the jungle, and Ugaki's plane going to the shoreline.*

What happened next during this historic mission has been the subject of several books and magazine articles, but all the official records, research by several noted historians and contemporary research all point to the same story. After the P-38s engaged the enemy aircraft, Lanphier did a climbing turn to the left (north by northwest) and went into a "head-on" duel with three Zero *fighters. By doing this, Captain Lanphier kept the enemy escorts busy and gave Barber a perfect opportunity to get a clear shot at his target, Admiral Yamamoto's aircraft, which was flying very low above the treetops.*

First Lieutenant Rex Barber then fired at the G4M1's right engine and next at vertical stabilizer of the aircraft. He could see bits of the engine cowling and rudder coming apart. After this, Barber shot once again at the right engine, which started to smoke very heavily, then showered the fuselage and left engine with his bullets. This spelled the demise of one of Japan's most influential and innovative leaders, Admiral Isoroku Yamamoto, his staff and all the airmen on the aircraft as it crashed into the jungle below.

The business end of Lockheed P-38G Lightning, "Miss Virginia." This was the fighter that First Lieutenant Rex Barber was flying during the historic mission when he shot down Admiral Yamamoto. While this P-38G was previously flown by First Lieutenant Robert Petit there were two Japanese "victory" flags and a ship representation painted on its nose for Petit's earlier claims in combat. (Jim Lansdale Collection)

(See pages 254-255 for aviation artist Roy Grinnell's painting "Mission Accomplished," which depicts First Lieutenant Rex Barber and his shoot-down of Admiral Yamamoto's aircraft.)

After First Lieutenant Rex Barber shot down Yamamoto's plane, three Zero *fighters chased after him and he did his best to lose them with some evasive maneuvering. He was very lucky because then two "friendly" P-38s (flown by First Lieutenant Besby Holmes and First Lieutenant Raymond Hine) came to his rescue and got the* Zeros *off his tail. During this skirmish, Holmes made claims for shooting two of the enemy fighters down and Hine shot at the third aircraft. After this Holmes and Hine, went after the second bomber (tail number 326) which carried Vice-Admiral Matome Ugaki. His aircraft was trying to flee away from the lethal P-38s by flying just above the water. Holmes proceeded to fire on the bomber and scored some hits, which silenced the enemy gunners along with some damage on the aircraft. First Lieutenant Holmes scored several hits on Ugaki's aircraft and stated that he shot it down. Barber also fired at this bomber so he, Holmes and Hine were responsible for "the kill." Nevertheless, whoever did the shooting; the results were the same, Ugaki's stricken aircraft crashed into the water below.*

Vice-Admiral Matome Ugaki survived the crash along with a two others on the aircraft; Captain Motoharu Kitamura, Chief Paymaster of Combined Fleet, and pilot Flight Petty Officer Hiroshi Hayashi. While I was researching in Japan, I located Hayashi, who was still healthy at that time. However; the veteran did not want to talk about his experiences during this mission. Ugaki survived until 15 August 1945 when he participated in a suicide air attack mission off Okinawa.

First Lieutenant Raymond Hine did not make it back from the mission and he was the only American loss for an otherwise "perfect" mission. Hine's fighter, which had a smoking left engine, was last seen by Holmes heading towards Wilson Strait, southwest of the Shortland Islands. All the other P-38 pilots returned back safely to Guadalcanal, but as Canning recalled earlier, First Lieutenant Besby Holmes arrived the following day after having to land his gas-starved P-38 fighter (on the uncompleted airfield in the Russells) then refueling there before heading home.

Left: First Lieutenant Raymond Hine was the only American pilot to lose his life during the 18 April 1943 mission to intercept and shoot down Admiral Yamamoto. (Second Yamamoto Mission Association via Jim Lansdale)

Right: The P-38 pilots who attacked the two Japanese bombers carrying Admirals Yamamoto and Ugaki. Left to Right, Captain Thomas Lanphier, Jr., First Lieutenant Besby Holmes and First Lieutenant Rex Barber.
(347th Fighter Group Association via Jim Lansdale)

Below: Surviving American participants of the Yamamoto Mission. The only man missing was First Lieutenant Raymond Hine who didn't return from this engagement. Left to Right, Back Row: Roger Ames, Lawrence Graebner, Thomas Lanphier, Jr., Delton Goerke, Julius Jacobson, Eldon Stratton and Albert Anglin. All aviators are first lieutenants, except Lanphier who was a captain at that time. Front Row: William Smith, Douglas Canning, Besby Holmes, Rex Barber, John Mitchell, Louis Kittel and Gordon Whittaker. All of these men are lieutenants except Mitchell and Kittel, who were majors at that time. (347th Fighter Group Association via Jim Lansdale)

Korea & Vietnam

Before World War II was over, you got points for all your awards and time/rank in the service. They wouldn't let me go on vacation during Christmas-time 1944, so I got out of the service. Nevertheless, I didn't like being out so I rejoined the Air Force. I flew P-51 Mustang fighters in the Korean War, which was a good aircraft in WW II, but not suitable to fly in this conflict. I flew thirty-seven P-51 Missions with the 40th Fighter Squadron, 35th Fighter Group. We lost over 100 Mustangs while I was in Korea. It was a good fighter but the aircraft wasn't suitable for air to ground strafing. The enemy also had jets so we couldn't dogfight with them. In fact, one Chinese down on the ground with a rifle could shoot down a P-51 by hitting the coolant system. When the coolant ran out you had just three minutes to get out of your stricken fighter. By pushing the primer button, you could get an additional fifteen seconds of fuel. If you were flying way up north and crashed, you would become a POW (Prisoner of War), but many of our guys didn't survive this ordeal.

Our losses in Korea were a lot worse than Guadalcanal. In my tent, five men were killed in combat, one guy got shot down three times and another one bailed out once. I got shot up on one mission, but made it back. Out of our two squadrons, the 39th and 40th, we lost twenty-five airmen and over 100 aircraft. It wasn't that the P-51 was a bad airplane; it just was the wrong one for the job. In retrospect, we should have used the P-47, which could take a lot of punishment, but still return home.

Next, I flew in a Weather Recce flight mostly to Russian and Chinese borders as they were not sending weather reports and somebody had to go look so Major Keohane and myself set up this outfit. During the Vietnam War, I flew C-133s out of Dover AFB (1st Military Airlift Squadron) mostly hauling helicopters and tanks over to Vietnam. Then, I had a desk job at Scott AFB from which I retired in late 1969 as a lieutenant colonel. After retiring from the Air Force, I moved to Florida because I liked it so much. I bought a house and lived here ever since. I worked as science teacher for twelve years until I retired in 1982 for a second time. I also enjoyed building RC models as hobby for many years.

Reflections

All together, I flew around 6,500 hours in the military. My Lord, I was shot at very frequently but actually only can recall being hit once in WW II and another time in Korea. I guess, I was a good pilot, but also lucky as hell. Flying for the Air Force was my job and I loved it. If I could do it again, I would. Most of my comrades had college degrees and this was because the Air Force wanted to see if you started something, you could finish it. In retrospect, during World War II, I thought it was my job to do and I did it. That's all there was to it, this was my job. So, it really was not personal and I had no animosity towards the Japanese. We killed some of their men and they killed some of our guys.

Doug Canning passed away on 20 February 2016.

[1] www.afhra.af.mil/

[2] These panels were used to identify the areas used by "friendly" troops and their enemy opponents. A few references that I read state the panels were a very vivid yellow color so they could be easily spotted from the air.

[3] Even though the belly (or drop) tanks were not the same weight due to being filled with different amounts of gasoline, they were mounted close enough to the center of gravity on the P-38 fighters that the tanks did not cause any performance-related concerns during this important mission.

[4] The majority of Japanese wartime aircraft didn't have self-sealing fuel tanks, armor or even excellent functioning radios. Japanese aircraft were typically designed with maneuverability, range and speed instead of "survivability."

[5] All of the background information and results of this historic mission were derived from three invaluable sources; historian Jim Lansdale's detailed analysis and two books, Carroll V. Glines, *Attack on Yamamoto* along with William Wolf, *13th Fighter Command in World War II, Air Combat over Guadalcanal and the Solomons.*

Chapter 2

"I didn't feel scared because I was supposed to sacrifice my life for Japan."

Warrant Officer Kenji Yanagiya (IJN)

WHILE I WAS LIVING in Japan, one of my objectives was to meet Zero *pilot Kenji Yanagiya. This amazing veteran was the only living Japanese survivor from the well-known mission that claimed the life of Admiral Yamamoto; a real-turning point in the Pacific War and an ominous sign of things that would be coming in the near future. When I met Yanagiya-san in his Tokyo home, he was very mentally sharp and articulate. During my visit, this distinguished aviator showed me some of his treasured mementoes from his Navy service including a flight log book, sword and signed flag.*

Early Years

I WAS BORN IN MARCH 1919 in Hokkaido. I had two sisters and two brothers. I was the oldest boy in our family. Then, our family moved and I grew up in Karafuto Prefecture (Sakhalin Island), a territory that was given to Japan after the Russo-Japanese War. This area was taken from Russia after they lost the war. Today, it is once again part of Russia.[1] My town had about 5,000 people and there was a pulp mill and coal mine. I lived there until I enlisted in the Navy. At that time there was a draft so as a young, healthy male you had to join either the Army or Navy. Most people joined the Army, but I preferred the latter service.

*****

IN JANUARY 1940, KENJI YANAGIYA joined the Imperial Japanese Navy at Yokosuka. After that, he graduated as a member of the Hei *3rd Class (Flight Training Course, 17th Class) in March 1942 at Ōita Air Base. Then, after graduation, Yanagiya-san transferred to the 6th* Kōkūtai *and went to Rabaul in October of that year. On 1 November 1942, his air group was renamed the 204* Kōkūtai, *which flew missions to attack Guadalcanal and did CAP (Combat Air Patrol) over ships going to reinforce the beleaguered forces on that island.[2] During this time, Japan was doing everything it could to win the Battle of Guadalcanal Island, and this became a big turning point in the Pacific War once Japanese forces evacuated the island in mid-January 1943. Yanagiya-san was credited with his first aerial victory on 5 January 1943 when P-38 fighters and B-17 bombers came to attack Buin. According to historian Henry Sakaida's research, two P-38 fighters (from the 339th Fighter Squadron) were shot down on that day.[3]*

Above: As a young Japanese naval aviator, Petty Officer Second Class Kenji Yanagiya. (Kenji Yanagiya Collection)

Right: An undated picture of Yanagiya-san in his flying gear. (Kenji Yanagiya Collection)

Yamamoto Mission

As mentioned in American veteran Doug Canning's previous account of the historic mission, on 18 April 1943, Admiral Yamamoto was shot down by P-38 fighters. Here is the Japanese side of the story. On that day, Kenji Yanagiya would have the duty of flying one of the six A6M3 (Model 32) Zero *fighters that were assigned to escort the two Mitsubishi bombers (which were converted into transports) that carried the famous admiral along with a number of other important Navy officers.*

At 0740, six A6M3s from the 204 Ku *took off from Rabaul to escort the two Mitsubishis (705* Ku*).*[4] *The two latter aircraft carried Admiral Isoroku Yamamoto and his staff in one plane, while the second bomber had Vice-Admiral Matome Ugaki and several dignitaries as passengers. The official* kōdōchōshos *stated that, at 0915, this group of two bombers and six* Zeros *engaged twenty-four P-38 fighters (but the actual total was sixteen, not twenty-four).*

Admiral Yamamoto would never return alive to Japan after his trip to the frontlines to help encourage his men during the I-Go Operation. In this image he is watching a Zero *fighter take off during his visit to Rabaul. (S. Bon Group)*

I WAS FLYING AT 3,000 meters (9,842 feet), which was 500 meters (1,640 feet) above the Mitsubishi G4M1 bombers. Normally, enemy fighters would attack us from above but this time, they were at 1,500 meters (4,921 feet) so it was easy for them to attack Admiral Yamamoto's aircraft. One moment later, I saw the admiral's bomber smoking and dropping into the jungle. Everyone on that aircraft perished. The second bomber crashed into the water and a few people survived. We were attacked by P-38 fighters and the commander was Major John Mitchell. After the enemy fighters shot down the two bombers, they returned to Guadalcanal Island. I tried to alert some of our men on the ground by firing my machine guns over the beach area to let them know something bad happened. I was later reprimanded for doing this because my actions were misunderstood. I was the only *Zero* pilot to chase after the P-38 fighters after they shot our bombers down. They were flying at normal speed and I shot one of them down (from behind). I saw fuel leaking out of this P-38 and the aircraft started to go down. After the war, I learned this pilot didn't survive.[5]

Since I was the only person to chase the P-38s, I was the last one to land at Buin Airfield (on Bougainville Island). Before I landed, the pilots from my group reported what happened. Only one of our *Zero* fighters had to land at another air base. Normally, the runway at Buin was very dusty but because Admiral Yamamoto was supposed to visit, they sprayed it with water so no dust came up when I landed. We really were shocked that enemy fighters came over this area to attack our group. Since I was the last person to land, I really couldn't see a lot of reactions from people on the ground. I did see a lot of higher-ranking officers waiting to see what happened. Then, I had to return to Rabaul.

A6M3 (Model 32) Zero *fighter, with clipped wings and a Sakae 21 engine. These modifications added some speed and improved the roll rate. Yanagiya-san was flying this version of the aircraft for his 18 April 1943 mission. (Kenji Yanagiya Collection)*

*****

After the huge loss of the two Mitsubishis and their important passengers, at 0945, one Zero landed at Ballale Island while the five others landed at Buin. However, out of the six surviving Zero pilots, four of them didn't even live to see 1944. As mentioned later in this account, both CPO Yoshimi Hidaka and PO1c Yasushi Okazaki died less than two months later on 7 June 1943 above the Russell Islands. Yanagiya-san was severely wounded on this same mission and never flew in combat again. Just nine days later, on 16 June, Lieutenant (junior grade) Takeshi Morizaki, perished in action by Lunga Point. Next to be lost was PO1c Toyomitsu Tsujinoue who failed to return from a mission on 1 July to Rendova.[6]

The sad remains of the Mitsubishi G4M1 bomber (manufacturer number 2656) which originally had T1-323 painted on the vertical tail. This is the aircraft that Admiral Yamamoto died in. Today, it still lies where it fell, near Moila Point on Bougainville Island. (Charles Darby via Jim Lansdale)

Feelings while flying missions

Normally, when I flew my combat missions, I didn't feel scared because I was supposed to sacrifice my life for Japan. However, I was a bit scared while landing after the Yamamoto Mission because visibility was bad and I couldn't land at my usual air base. So, I had to land at a different air base and landed very hard on the runway. I really didn't feel scared during my missions but I felt it was difficult to sometimes deal with unplanned situations while flying.

Yanagiya's last combat mission

Kenji Yanagiya's last combat mission occurred on 7 June 1943 while doing a sweep over the allied air bases in the Russell Islands. On this date there was a lot of fierce aerial action over the Russells and on the Japanese side, three kōkūtai *were involved: the 204, 251 and the 581. I will include the actions of each group separately so the aerial actions are a bit easier to comprehend. All of the background material (of the Japanese action on this date) was derived from the* kōdōchōshos *(the official Japanese war records) of each unit for historical accuracy.*[7]

Yanagiya's unit, the 204 Ku, *consisting of twenty-four* Zero *fighters took off from Buin at 0917. During June*

1943, the 204 Ku *used Rabaul as their main base (twenty-one sorties) but flew missions from Buin on 7th, 12th and 30th of June. Whenever this air group was assigned to fly as far as places such as Rendova Island, the Russell Islands and Guadalcanal Island, they went to Buin, the day before the mission, then flew from this airfield the following day.*

For this particular mission on 7 June 1943, about one hour into the flight, eight of the Zeros *engaged the same amount of Grumman F4F Wildcat fighters, then at 1045, Yanagiya's group attacked their target using incendiary bombs. After completing their attack, the aircraft of the 204* Ku *were once again engaged by estimated thirty enemy fighters including P-38s, P-39s and P-40s. According to Japanese accounts, this second engagement lasted about twenty minutes. Then on the way home, there was a brief third contact made with enemy aircraft but it is not clear if any (on either side) were shot down at that time. After this, the surviving* Zero *fighters returned to Buin at 1250 and later flew back to Rabaul.*

A great perspective of both versions of the A6M3 Zero *fighter during flight operations on Rabaul. The first aircraft is the Model 32, the clipped wing-version, followed by a Model 22, with rounded wing tips. (S. Bon Group)*

On the same day (7 June 1943), at 0845, the 251 Ku *launched thirty-six Type 0 carrier fighters from Rabaul bound for the Russell Islands. As a side-note, this group during June 1943, utilized three bases for their missions: Rabaul, Buka and Ballale. On this mission, they arrived over their "target" area around 1110 but were intercepted by an estimated 100 enemy aircraft between 1130-1150. After this aerial action, the surviving* Zero *fighters flew to Buka and arrived there at 1430 then returned to Rabaul about two hours later. As for the 581* Ku, *on that same day, at 0915, twenty-one Type 0 carrier fighters took off from Buin bound for the Russells. Six Zero fighters (of the Third* Chutai*) engaged four Grumman fighters at 1010 over Gatukai Island but the rest of the Japanese fighters didn't participate in this action. The main body of Type 0 fighters continued on and arrived at the Russell Islands at 1045. They headed back around one hour later and landed on Buin at 1330.*

*****

DURING A MISSION TO attack an airfield on the Russell Islands, my *Zero* fighter was armed with two *Shoi-dan* (incendiary bombs) bombs and flying at 8,000 meters (26,246 feet). When I went down to 6,000 meters (19,685 feet), I released my bombs and started to climb. Then, all of a sudden, I was attacked by enemy Grumman F4F fighters from the rear (of my aircraft). My right hand, which was holding the control stick, was obliterated by .50 caliber machine gun shells. Even though there wasn't much left of my hand, my tattered right glove was still on the control stick. My leg was also wounded but my *Zero* fighter was okay. I had to escape by flying at 100 meters (328 feet) so it would be difficult for enemy fighters to chase me. I couldn't fly my fighter very well and was in great pain. When I tried to land on Munda Airstrip (New Georgia Island), I had to fly over a twenty-meter (sixty-five foot) palm tree, which was a bit unnerving but I still managed to land safely.

After I landed, a doctor treated me but I really wanted to get back to Rabaul as soon as possible. Since the doctors were short of medical supplies, there was no anesthesia to use during my first operation to remove the remains of my tattered right hand. Three people had to hold me down during an operation and it was very painful. I remember very clearly seeing my nerve being pulled out. Because this air base was near the front-lines, I couldn't travel during the day so I had to always move at night by small boat and it took seven days to get back to Rabaul. My arm was rotting away and flies were flying around it. Once I returned to my air base, I had two more surgeries. All together, I needed three surgeries to heal my wounds. While I was in the hospital, I met my future wife, who was a Red Cross nurse.

After I recovered from my wounds, I never flew combat missions again. When I first went to Rabaul, Japan was winning the war but gradually the momentum shifted and the United States became stronger than our side. By the time, I left Rabaul, our side was much weaker.

*****

According to the postwar research by Henry Sakaida, on this date, only one Grumman Wildcat fighter unit engaged Zeros, *which was VF-11, the Sundowners, of the United States Navy. From this group, two pilots put in claims for* Zero *fighters. Between 1010-1035, Lieutenant Commander Gordon Duward Cady claimed one* Zero *and a second probable over Vanganu Island. The other Navy aviator, Lieutenant Raymond W. Vobel Jr. a* Zero *between 1130-1134 over Baruku Island.*[8]

The Imperial Japanese Navy lost a number of airmen during the 7 June 1943 mission to attack the Russell Islands. Members of the 204 Ku *who perished that day were CPO Yoshimi Hidaka, PO1c Yasushi Okazaki and PO2c Kameji Yamane. Two additional members of this group (not including Yanagiya-san) also landed at other airstrips after the mission, PO2c Yanagitani landed on Munda while CPO Hideo Watanabe went to Kolombangara. This group claimed eight allied fighters shot down and one more probable. The airman also thought they scored one bomb hit on an airstrip in the Russell Islands during the mission.*

Five pilots of the 251 Ku *also didn't return from this mission including PO1c Masuaki Endō, PO2c Toshitaro Sekiguchi, PO2c Kanichi Masuda, PO2c Setsu Matsuyoshi and PO2c Yutaka Fukano. This air group made claims of eighteen allied aircraft shot down and five more probables. The 581* Ku *was fortunate and lost no men that day.*[9] *They put in claims of having shot down four Grumman fighters that day.*

Yanagiya's last days in the Navy

I became a flight instructor in the fall of 1944 and did this duty until the end of the war. During the war, I attacked eighteen enemy aircraft, but I'm not sure how many aircraft I shot down. After the war, I tried to hide my wartime flying records because I didn't want to be considered a "war criminal."

*****

Yanagiya's last rank in the Navy was warrant officer and he ended the war with at least eight victories.[10] *He was a fortunate survivor of several dangerous engagements during the war and to put this in perspective, out of the 101 airmen in his pilot training class, only nine others survived the war.*[11] *Per his wartime flight log book, at the end of April 1943, he flew 493 hours.*

Left: Yanagiya-san in his flight gear towards the end of the war after recovering from his wounds. (Kenji Yanagiya via Henry Sakaida)

Above: Warrant officer Kenji Yanagiya while serving as a flight instructor. He is standing next to a Kyūshū K11W Shiragiku *training aircraft. (Kenji Yanagiya Collection)*

Feelings Today

The peace that Japanese people have today is wonderful. I had a good experience visiting the United States to attend the symposium at the National Museum of the Pacific War. The American veterans and other people I met welcomed me very warmly. Personally, I believe some younger people like my grandson understand what happened in the war.

Yanagiya-san died on 29 February 2008 at the age of eighty-eight at his Tokyo home.

Kenji Yanagiya during the author's 2002 interview with him. (Ron Werneth)

[1] As Yanagiya-san stated Karafuto Prefecture was once a part of Russia and given to Japan as reparations after the Russo-Japanese War. This prefecture was located on the southern half of Sakhalin Island, but Russia still retained the northern portion of this island after the above conflict. It is located between the countries of modern day Japan and Russia. In 1952, Japanese gave up its claims for this territory but this issue has been a "sore spot" on the relationship between the two countries to this day.

[2, 6, 9 & 11] Ikuhiko Hato and Yasuho Izawa, *Japanese Naval Aces and Fighter Units of WW II.*

[3] Henry Sakaida, *Imperial Japanese Navy Aces 1937-45*. Hereafter, *Imperial Japanese Navy Aces 1937-45.*

[4] Unfortunately, while researching this engagement, we saw that the original 705 *Kōkūtai kōdōchōsho* for this important mission was torn out of this unit's original flight records. It may have been stolen or destroyed in the postwar years. So, we couldn't "confirm" the exact takeoff time, etc. for the two Mitsubishi bombers that may have been recorded for them.

[5] Yanagiya-san may be talking about First Lieutenant Raymond Hine, who is mentioned in veteran Doug Canning's story in this book.

[7] This information was kindly provided by historian Minoru Kamada.

[8] www.j-aircraft.org/

[10] *Imperial Japanese Navy Aces 1937-45.*

Chapter 3

"There hardly is a day when I don't think of Big Boy."

PFC Ed Adamski (USMC)

I FEEL VERY FORTUNATE to have met WW II United States Marine Corps veteran Ed Adamski. He is one of the last surviving K-9 (scout dog) handlers who fought in combat during the Pacific War. Ed is a local Chicago-born veteran and now lives in nearby Schererville, Indiana. You may wonder why a book filled with mostly aviator stories has an account of a Marine that fought with a dog in land combat. I purposely wanted to include Ed's story because he was one of the "grunts" on the ground and they worked very closely with their airborne counterparts to help get the job done. Furthermore, people need to know about how brutal land-based combat was both to human beings and animals in the Pacific Theatre.

Ed's story when he served with the 2nd War Dog Platoon is an amazing one. He fought "his" war with a four-legged companion named "Big Boy." This Marine's partner may not have been a show dog; however, this did not matter when it came to fighting against Japanese soldiers who were willing to fight to the death. During our conversations together, he spoke very frankly about his wartime experiences and wanted to make sure that future generations will know what really happened. Ed was very humble, but proud, of what he and his Doberman did during the Battle for Guam. This is their story.

Early Years

I WAS BORN IN JULY 1924 and lived in the Chicago south-shore area on 83rd Street. My community was up against the southern end of Lake Michigan. I joined the United States Marine Corps in March 1943. I enlisted because I was going to get drafted and wanted to select the military branch (of my choice). I did not want to go into the Army. I chose the Marine Corps because there were a lot of stories of the early battles in the Pacific and they sparked my interest. I was eighteen years old and gung ho. (Later on in my life), when my sons were enlisting in the military, I insisted that they stay out of the Army or the Marines. This is because I learned a "hard lesson" (as a Marine myself), that you never had a clean bed or three square meals. But, in the Navy (or Air Force) you did.

Marine Corps Training Days

AS I WAS JUST finishing up boot camp, a drill instructor said that they were organizing a new unit with war dogs.

It sounded interesting and they wanted volunteers so I volunteered with some others. We were on the west coast at that time and had to go to North Carolina. There, outside of the main base at Camp Lejeune was an ex-Civilian Conservation Corps (CCC) camp from the depression years. It was right on the river between the town of Jacksonville and main base. So, the area was fairly isolated.

We went through vigorous training with the dogs. This training started in May 1943 and continued to the end of November. Then, we boarded trains to go to the west coast. There we camped in an isolated canyon, outside of San Clemente (in California). We were assigned to a Marine Raider battalion and trained with them. They were laughing and barking like dogs. The Raiders thought we were a joke, but they found out differently. As a matter of fact, we even outdid them on these forced marches to high peaks. Our dogs did a hell of job too. For example, during training, the dogs searched out the "enemy" and they could never infiltrate our lines. Initially, we were supposed to go overseas with the Raiders and do special missions. This is what our 1st War Dog Platoon did.[1] However, the order came down to disband the Raider units (so we never went into combat with them).

Big Boy

NINETY PERCENT OF our dogs were Dobermans donated by the members of the Doberman Pinscher Club of America. The other dogs were mixed breeds. Big Boy was a Doberman and I'm not sure how he was assigned to me. However, his original owner Charles Jackley, was from Chicago, and I was from Chicago so that might have had something to do with it. Each dog had a number tattooed inside their ear for identification, Big Boy was number 98. They gave the dog handlers the owners' addresses and we wrote them and kept them informed about how the dogs were doing. Big Boy was not a show dog, his ears were not cropped and he had a defect in his hind quarters that gave him a poor gait. I think this is probably why the owner donated him. He was probably one of the poorest looking dogs in the litter but Big Boy was very intelligent. He seemed to pick up things quicker than the other dogs and never failed to obey a command. I never had a problem with him. Everybody used to laugh at him because of his appearance and gait. However, later they learned he was a sharp dog and I felt very fortunate to have him. Big Boy was very dependable and could "sense" my mood and I could "feel" him picking that up. Like many of us the dog suffered the same hardships just like we had. When these dogs were transported, they were in individual kennels with not a lot of room. It was very difficult on them.

Above left: A young Ed Adamski in his Marine dress blues. (Ed Adamski Collection)

Left: A rare unpublished picture of Big Boy, which was taken from his official Marine Dog Service Record Book. Note that he is standing in front of a wall with 98 on it, which was his serial number. (NA)

Life in a War Dog Platoon

We were a small unit and seemed to be on the tail end all the time and always got the short end of the stick. We felt that we were treated like crap by the Marine Corps and this could be for anything, from promotions to even having a hell of a time getting good food. Unfortunately, our officers were not ready to fight for us. To be a dog handler you could only be a PFC (private first class) and if we were promoted to corporal, we would lose our dogs. We should have been promoted more, but we were not. Our Platoon Sergeant, Ivan Hamilton (3rd War Dog Platoon) was an old Corps guy who enlisted years ago and was a "lifer." He tried to get our commanding officer to talk with a colonel that he (Hamilton) served under years ago for additional promotions. It is hard to knock our own guys but we lost out because we did not have the brass to fight for us.

Going Overseas

We shipped out on 24 February 1944 aboard a merchant ship called the SS *Skinner* (ID-1321). It took thirty-eight days to go overseas because we could not travel in a straight line. We were headed for Guadalcanal but on the way, we stopped in a harbor at New Caledonia. Our officers were going ashore but before they left, there were orders for no swimming or diving in the harbor, but they did not tell us why. Well, we were young and stupid and we started diving off a shower building on the deck. This building was about ten feet high, so I would say we were diving at least forty feet. I decided to go up and take a dive and as I was going down, I did not want to dive too deep so I arched my back and hit the water with my chest. My heels came over and hit the back of my head and I thought my back was broken but I did not complain. When they, our officers returned, they said, "You damn fools, there are sharks in these waters." We pulled out of New Caledonia after two nights and headed for Guadalcanal.

Ed and his platoon arrived on Guadalcanal on 1 April 1944, a Saturday.[2]

This island was one of the worst that I ever was on. It was nothing but jungle. You could take a shower every ten minutes but you would not feel clean or dry. Fungus was all around us and guys would have their hair falling out. Some of the men got a disease called elephantiasis

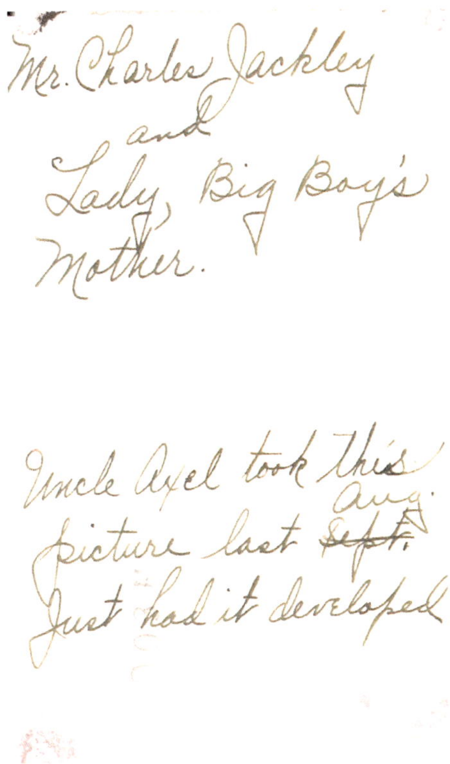

Photo, captioned on the back, of Doberman Pinscher breeder Charles Jackley, with Lady, Big Boy's mother. Remarkably, both Adamski and Big Boy came from Chicago. (Ed Adamski Collection)

which would cause their limbs to swell.[3] I even saw a native with testicles so swollen; they were in a little wooden wheel barrow.

When I first got there, I tried to save myself some time by putting my dungarees in a bucket of soapy water instead of scrubbing them. Unfortunately, soon after when I went to check them, the bucket was already full of maggots. Mosquitoes and malaria were very bad and I don't think one person got out there without malaria. We stayed there for about three months going through training maneuvers and doing simulated landings. We were held in reserve for the invasion on Saipan. After about three days there, they let us go to the Marshall Islands. There were mostly atolls there. We did some training but it was very rough on the dogs. They were thirsty and we didn't have much water. Therefore the dogs started lapping up sea water which was very bad. We had a heck of a time trying to prevent them from doing this. They kept us out there all day long. It was very miserable, especially for the dogs. Then, we headed off to Guam.

This picture was taken in May 1944 on Guadalcanal. All of these Marines hailed from the Chicago area. Left to right: William Harper with Laddie, Joseph Schwartz and Pal, Ed Adamski with Big Boy and Charles Peak with Hans. All of the men should be PFC in rank, during this time period. (Ed Adamski Collection)

*****

ACCORDING TO THE ALL the wartime documents in Ed's United States Marine Corps Service Jacket, he embarked on the USS Titania *on 1 June 1944 and sailed on this same date. Seven days later, the ship arrived at Kwajalein, Marshall Islands and stayed there until the 12th of that month. Then, on 28 June, the* Titania *arrived at Eniwetok, Marshall Islands. After this, on 17 July, they sailed from Eniwetok and arrived at Guam, Mariana Islands, four days later, and took part in the combat operations there.*

Combat on Guam

THE ISLAND OF GUAM had attributes that made it an excellent location for future military operations against the Japanese which included two airfields and a harbor. The U.S. Army Air Force could operate their "bludgeon" of the Pacific Air War, the Boeing B-29 Superfortress from these airfields while Apra Harbor was deep enough to accommodate large allied ships. The Japanese had occupied the island since 10 December 1941 and they would not give it up without a fight.

At 0530, on 21 July 1944, the liberation of this island officially began with an intense two-hour naval bombardment by the large armada of American ships offshore. However, prior to this day, for almost a month, the Japanese troops and defensive positions were under constant attack by both ships and aircraft. This continued on the day of the invasion when aircraft supporting the landings roared in and tried to help soften up the enemy's fortifications. At 0829, the first wave of troops from the Third Marine Division landed on the beach area between Asan and Adelup points, which were north of the Orote Peninsula. Shortly, after this, the 1st Provisional Marine Brigade began their landings in a region south of this peninsula near the village of Agat. After these two initial Marine landings, men of the Army's 77th Infantry Division were later called in to help liberate the island.[4] While on Guam, Ed saw heavy action with his K-9 partner Big Boy and remarkably survived two banzai (suicide) attacks.

Map depicting the progress of the Battle of Guam (1944) from the beach landings until the successful completion of the campaign. (United States Army)

This is how Ed and Big Boy were lowered to their landing craft during the Guam campaign. Not easy to do, especially with a dog. This picture was taken a year earlier during the landings on Bougainville Island, but is a good depiction of what Ed and his dog did during their Guam landing. (USMC)

Our sixty dogs were divided between different companies of the 3rd, 9th and 21st regiments of the Third Marine Division. I was attached to F Company, 9th Marines that would land on Guam. We spent two days offshore and we were really hitting it with everything we had. One of the guys said, "My God is there anybody going to be left for us to kill?" Our troop transport had cargo nets strung on the side of the ship and Higgins boats would be used as our landing craft. We would get one or two guys to climb down in the landing craft and then lower our dogs while we climbed down the nets. Boy, the Higgins Boat and the big ship were bouncing around a lot so it was pretty dangerous. We went to a coral reef where there were supposed to be half-tracks to take us the rest of the way in. Well, they must have been knocked out because there weren't enough of them (for us). Shells were hitting the water around us and some guys in the landing crafts were blown right out, but nobody from my unit was lost (at this time).

The ramp on the Higgins Boat dropped and we stepped off. I was in water up to my armpits and weighed down with a cartridge belt with .30 caliber ammo, M1 rifle and a double pack on my back with extra clothing and food (for both me and my dog). The dogs had no life vests so Big Boy started swimming away. I stepped in a shell hole and was going down and wasn't coming up. One of the guys pulled me up by my helmet and I was really lucky to have had my helmet buckled up. You see some of the war movies where guys don't have their helmets buckled, but this never happened in real combat. The helmet won't do a hell of a lot of good if it wasn't buckled up. This happened while we were wadding in so I almost never made it to the beach.

When I finally got on the beach there were guys lying behind a berm of sand. I tried to dive between them but the guy on the left would not move over. I looked over and saw that his whole backside was blown away so he was dead. That was my first experience with something like that.

Actual photo taken by an unknown Chicago Tribune *correspondent during the landings on Guam. According to Ed, even though Big Boy looks serene and the picture seems "posed" – it wasn't. The veteran and his dog were under fire while this picture was taken so it wasn't a safe place to be. (Ed Adamski Collection)*

Third Marine Division K-9 handlers and their dogs advance as they move forward from the northern beachhead. (USMC)

We moved up about fifty yards inland and I started digging my first hole. This guy comes out of nowhere and takes my picture and I thought that he was nuts because shells were dropping all around us. This guy (photographer) asked me where I was from and I said Chicago. He replied that there was a *Chicago Tribune* correspondent there and he would send him over but, I never did see this correspondent again.

Three of my buddies got hit right there. Two corpsmen were carrying PFC Raymond Rosinki back to the beach and I asked how he was doing and he replied, "Ski, don't worry, just a nick." I found out later they buried him at sea. He was hurt a lot more than he admitted. They were dropping mortar shells like mad in that area and I never did get that hole finished. The next thing I knew was we had an order to move up to a coral ridge about 30-40 feet high. I went up along the ridge with a reconnaissance man and some others.

That night they were hammering that ridge with mortar shells, which seemed to start on our right flank just below the top of the ridge. After this, they would move it up a notch and come closer. It was the same pattern, over and over. I will tell you right now, I heard two voices (during this time). One guy was saving me while the other one was taking my life and making a mockery of it. Normally, I only talk about this stuff at reunions with the guys who know what the hell I am talking about. But I am hoping that some good comes out of this and people can learn something about war. I also want to leave a legacy for my kids and grandchildren. So I am being as open as I possibly can. I was scared stiff and I could hear those voices as loud as I can hear my voice right now. We could not dig a hole because it was solid coral rock and we were out in the open with no cover. This was going on all night long and it seemed if they moved one more notch, we were going to get it. I was with another guy and there were similar pairs along the line who were maybe five yards apart from the following pair. We were on the outer perimeter of the whole troop landing. My dog was lying alongside me all night long and was calm as hell and acting like nothing was going on. I told him to stay, sit or go down and that is what he did. PFC Ed Topka, a guy from my unit, was killed that night.[5] We got off the ridge and went to the right where there was a narrow slightly paved road with only enough room for one vehicle to go by. We went down this path to a ravine where I'm sure there were Japs on the other side with their mortars.

On the second night, we dug in at the mouth of the ravine. My reconnaissance man was a scout, he strictly did patrols. He had been on Bougainville so he had some experience but he had never experienced being on the perimeter at night. Together we were assigned with a .30 caliber machine gunner on the outer perimeter. I could hear the Japs who were in supply caves on both sides of the ravine and they were working the bolts of their rifles back and forth all night long trying to draw your fire. I kept telling our guys that we should shut them up but they said that we could not do anything because if we did everyone else would start firing and start a war. Finally, the guys told me to go ahead. So I let off a whole clip (eight rounds) and Jesus, everyone down the whole line started firing thinking the Japs were coming in. The next morning Captain Wilson wanted to know who the hell started a war. My reconnaissance man said that he would snitch on me but never did and neither did the machine gunner. In 1992 I finally "confessed" to Captain Wilson and told him about it. I was on the point that night and he (Wilson) was fifty yards behind me when this action occurred. Louis H. Wilson, Jr. later became a four-star general and Commandant of the Marine Corps. He was awarded the Medal of Honor for his actions in combat on Guam.

Captain Louis H. Wilson, Jr. was awarded the Medal of Honor for his actions during the Battle of Guam. During this respective engagement, on 25-26 July 1944, he was in command of a rifle company attached to the 2nd Battalion, 9th Marines, Third Marine Division. While leading an intense ten-hour battle against Japanese Army troops on Fonte Ridge, he was wounded three times but never stopped fighting. During the struggle to capture this strategic high-ground an estimated 350 enemy soldiers were killed. After being wounded, Wilson was evacuated to the U.S. Naval Hospital in San Diego and remained there until 16 October 1944. (USMC)

Members of the 2nd Marine War Dog Platoon who served with a reconnaissance unit during the Battle of Guam. In the past, Ed said that could identify them by name but unfortunately, now that isn't possible now because of his deteriorating eyesight. (Ed Adamski Collection)

On the third day of the invasion, Captain Wilson called 20-25 men together. There were four dog handlers in this group including myself who were trained in scout work and were primarily on the point of a patrol. On this patrol, we had two K-9 handlers on point, as flank guards. I was in the right flank position with Big Boy and another man with his dog was on the other side. We also had two K-9 teams (one on each side) in the rear of the group. Before we left on this patrol, Captain Wilson said that the Navy was informed that we would be in the area so it would not be shelled or strafed by aircraft. We were given three objectives. One was the power station, which was knocked out. Second was an enemy-held warehouse which held some vehicles and supplies. Lastly, the third objective was to make physical contact with the 1st Marine Brigade who was coming in on the opposite end of the island. We made radio contact with the 1st Marine Brigade but just before we made physical contact, the Japs on the peninsula started shelling us so we could not come any closer. We were going through this area without any prior knowledge that the Japanese had dug pits and put dummy artillery guns in them surrounded with land-mines. They probably thought that we were the main force coming through. Well, we thought that the Navy was shelling us but it was actually the Japs! We had a hell of time getting out of there. I'd swear to God while I was running my body was lifting a foot in the air (due to the concussions)! Luckily, I don't think anyone was killed and we just ran back to our lines.

That night, we enjoyed some *saké* and canned crab meat (captured from the Japanese). Some of the guys found this stuff in a Japanese warehouse and were passing it around. This was the most delicious crab meat that I ever tasted in my life! As a matter of fact, it was the first food I had since leaving the ship and this was my third day (on Guam). All I did was drink water. The same thing went for the dog. I was so damn scared that I did not have time to eat. You only worried about the how the hell you would stay alive.

On the fourth day, we went up to Fonte Ridge right below Mount Tenjo. On the night of 24 July, it was raining like hell and my buddy and I were the rear guards with our dogs. Both of us dug foxholes and were laying there trying to keep our heads out of the water. All night long we were shooting up flares so we could see any infiltrations in the area below Mount Tenjo. It was fairly quiet. The dirt was red, the of the iron ore and in the morning when we got up, we were half red and half green. We looked like clowns.

The next day they pulled me away from my buddy, Pat Dale, and sent me to the perimeter further down the line with a .30 caliber machine gunner. It started to get dark and I was looking below the ridge (near Mount Tenjo), over to our left. There was a clearing and I could see movement and recognized the Japs. I said to the machine gunner, "Down there, look!" He could not see them so I loaded some tracers in my rifle and fired at the spot. This was the first warning that the Japs getting ready to move in. They hit us with a *banzai* (suicide) attack. That night Big Boy alerted two or three times when there was some movement. We had to learn the signs of our dogs and how they alerted and Big Boy alerted by doing a low-growl and the hair on the back of his neck would stand up. On this particular occasion, he was lying down but he would rise up a bit and you could see his hair stand up. Most of the dogs use the same method but some dogs were different. I told the machine gunner and he passed it on to the sergeant who would communicate by radio to the Command Post that was behind us. The Command Post would direct artillery fire, etc. (on the target area).

They hit us that night (the 25th), and the reason I remember this so well is because 25 July was my 20th birthday! We got hit again the next night (the 26th) and to be honest with you, I was pretty much out of it because everything was so God darn chaotic. The Japs were right on top of us and they were screaming and firing at the same time. They used everything from pole charges to throwing hand grenades and it was one hell of a mess. Maybe some of our own guys were hit by American fire. I got hit by grenade fragments and was knocked out.

On the morning of the 27th, I woke up on a stretcher in a field hospital, which was back in the ravine area. Evidently, I was unconscious for some time because I did not even remember how I got there. Prior to being wounded and taken to the hospital, the Japs tried to get in the ravine to some of their supplies from the caves. When they got there, they found some United States Navy doctors and corpsmen that were setting up this hospital. It was a real chaotic situation. These poor buggers never expected to be involved in combat, but in the Pacific Theatre, there were no rear echelon areas (that were safe).[6]

This is the actual field hospital that Ed describes in his account. Just prior to the young Marine being wounded and arriving there, doctors, corpsmen and patients had to fight for their lives during a Japanese banzai *attack. During this attack, one patient was wounded and a medical officer and a corpsman died. An additional thirty-one other men includes Marines, corpsmen and officers were wounded. (USMC)*

Once I went to the field hospital I lost Big Boy. Our guys were probably afraid to handle my dog so Big Boy just took off looking for me. Later I found out that he was out there for two or three nights down near the beach howling. The men were going to shoot him but, thankfully, they finally got him in. I have no idea how they did it. I was at the field hospital for three days but due to the chaotic situation, I was never registered. Evidently, the corpsman (on the battlefield) took care of my wounds and must have taken out a lot of shrapnel from my chest. I was also suffering from malaria and dysentery and was so damn weak I could not even stand up. One night a corpsman came in with a big gallon jug half-filled with capsules saying "Help yourself." Later, we found out these were sleeping pills and this was all the treatment we got. They were scared of being hit by the Japs that they were not even taking care of people like me. The only thing I could do was take the liner from my helmet and roll off the stretcher and pass blood and water (into it). I don't remember them feeding us so you know how chaotic it must have been. My wounds were superficial so after three nights, they let me go. The next thing I know is I joined a unit that a buddy from Gary, Indiana was in and he tried to help me. I was going through a malaria attack again and had a fever and was shaking like a dog and still getting soaking wet. Just before dark, a Jap ran in amongst us, with his hands in the air, bare-chested with nothing on his feet and surrendered.

The next morning my buddy told one of the officers (in this unit) about my condition and he said I should go down the road to Agana and find a field hospital. Nobody was with me and I went down this road by myself, sicker than a dog but I found a field hospital in a bombed-out building. They put me on a stretcher (on the floor) and took my temperature. Then, the next thing you know, there was some shelling and I jumped out the window. Back then, they called this being "shell shocked." Today, they call it PTSD. In those days, you would never mention it to a Navy doctor. The attitude was, "Suck it up." Everyone had the same problem. Even in the VA hospitals after World War II, you did not bring it up because they would call you a whiner.

Here I am in this condition and I joined up with another unit and I was helping a BAR (Browning Automatic Rifle) man carrying his rifle because he had a lot of ammo. All I could think about was keeping myself alive and pushing along with the main force and going the whole length of the island. I was with other dog guys

A clearly staged but unique photo that shows an unnamed Marine Doberman Pinscher resting under a camouflaged USMC poncho and surrounded by the typical gear that was used in battle. This picture was taken on Guam but no other details are known. (USMC)

but they were in a totally different unit. They actually should have sent me to the rear and back to my unit where I could I find out where my dog was. I ended up fighting with the 3rd Marine Regiment during their push across the length of the island (from 30 July-10 August). They declared the island secure but this was bullshit because thousands of Japanese troops were still hiding in the jungle.

We set up our camp on the island, but we had to go on patrols everyday to get these buggers out and luckily we did not lose any men or dogs during that period. On Saipan, where half our guys were, they also did these patrols and it was pretty dangerous work because every day you were going out and putting your butt on the line. I finally went back to the beach where I found the rest of the war dog guys. They did not know where I was and thought I was dead. But, they had Big Boy and we were together again.

*****

During the recapture of Guam, the Marine Corps (including Navy personnel who were assigned to the respective Marine units for combat) had 1,190 men killed in action, 377 died as a result of combat wounds and 5,308 were wounded. Their counterparts in the Army, the 77th Division lost 177 men and had 662 wounded.[7] As for the Japanese forces, they had 18,377 soldiers killed and 1,250 of their men became prisoners of war.[8] Remarkably, over 8,500 Japanese were killed or captured <u>after</u> the island was technically in American hands on 11 August 1944.[9] These surviving enemy troops kept Ed and the other K-9 handlers "busy" as they continued doing patrols to locate them until the end of war one year later.

The Second and Third War Dog Platoons lost three men and twenty-five dogs during the fierce combat on Guam. Those who visit the now-peaceful island can pay their respects to the fallen canines at the War Dog Cemetery located on the U.S. Navy Base at Orote Point. This cemetery was dedicated on 21 July 1994 due to the efforts of veterans and war dog supporters. A stone memorial adorned with a noble looking Doberman Pinscher is inscribed with the names of these often forgotten four-legged heroes that died in combat alongside the troops they served.[10]

Above: The original Marine "War Dog" cemetery on Guam. During the postwar years it fell into disrepair but was later properly restored and a monument was added to honor the heroic USMC dogs and their handlers. (Ed Adamski Collection)

Right: The modern day War Dog Memorial on Guam is visited by United States Navy Petty Officer 2nd Class Blake Soller and Military Working Dog (MWD) Rico as they pay their respects to fallen dogs that lost their lives in that battle. (USN, photo taken by Petty Officer 2nd Class John F. Looney)

Daily life as a K-9 handler

DURING THE FIGHTING ON Guam, the company commanders were using us double-duty. In the daytime, we would go on these patrols, and at night, they would keep us working an outpost or perimeter position. The latter duty was typically done with a .30 caliber machine gunner or radioman. We were never trained to do perimeter guard duty but we had to stay awake both during the day and night, to make sure if the dog alerted to something we were awake to realize it. The other guys (such as the machine gunner or radioman) could sleep because they depended on us. None of the officers ever took our names so they did not know who we were. They used us in situations that they should have used their own men. While ferreting out the Japs, once the dog alerted, we should have been pulled back and their guys should have gone in. Instead, they had us going in further which was how we lost most of our guys on Iwo Jima. Everyone knew the Japs were there so there was no point for us to go in. It seemed like they were "protecting" their guys and sending out the guy and his dog that they didn't know. That really upset me and I got even angrier thinking about it later.

We all suffered because of the combat and Big Boy suffered a lot. Since I was scared, I was not even thinking of feeding myself, therefore, I was not even thinking of feeding him either. But, we still got through it. I did not even get my shoes off for a couple of weeks so you can imagine what my feet looked like once they came off. I was peeling the flesh off my feet. You were soaked from going through a lot of rain. You did not bathe; there was no time for that. Furthermore, there were no facilities for bathing and there was not enough water to do that. You barely had enough water to drink. Normally, if you ran out of water, you would fill up your canteen from a creek or some rain puddles. After you filled your canteen, you would throw in a chlorine tablet then shake it up. A lot of the guys got dysentery from that because of the dead bodies. Those bodies would burst and there would be maggots along with blowflies. There were flamethrowers that burnt a lot of them up so it stunk like hell. When they burned them, a long trench was dug out and the Jap bodies would be shoved in like garbage and covered up. It was pretty crappy.

The End of the War

AFTER GUAM WAS SECURE, I stayed on this island and worked with the Third War Dog Platoon while the other half of my platoon went to Saipan. The Navy moved their operations from Hawaii to Guam which ended up being the main base for other operations such as Iwo Jima. We set up our camp above what they call Tumon Bay and on the weekends we enjoyed some recreation such as baseball.[11] During the week, we were on patrols and we were finding Japanese for months (after it was declared secure). They were all over the island in the jungle and caves and would not surrender.

After the major combat subsided on Guam, Ed stayed on the island for cleanup operations for Japanese stragglers. But, during this time, it wasn't all work. He also did some competitive boxing with the USMC Boxing Team. It is not known who is pictured with the young boxer. (Ed Adamski Collection)

June 1945, a group of some of the Marines that were left on Guam to complete all mop up operations and clear out the remaining stragglers. Ed is crouching in the first row, far right. (Ed Adamski Collection)

Return Home - 1945

Ed left Guam on 28 October 1945, he and his dog were finally going home. They arrived in Norfolk, Virginia on 6 December and Ed received his discharged twelve days later.

*****

I stayed with Big Boy until the end of the war and afterwards we didn't have a lot to do with the dogs because they could evaluate them fairly easily. Dobermans are very intelligent, sharp and train well and are always watching everything carefully. In general, a Doberman will do what he is trained to do. They like to be commanded by one individual but will take commands from others. Big Boy had a good demeanor and there wasn't a mean bone in his body but if you wanted him to attack, he would attack viscously. During wartime training we had guys dressed in padded suits and had our dogs attack them. We also tied them to a spring lead on the wall and had a guy come along who agitated the dogs. This drove them crazy. In hindsight, I wonder if that was a wise thing to do. It occurred because the people who were helping with the training of the dogs were not knowledgeable about combat.

When I went back (to the United States mainland), I had to send Big Boy back to the original owner who had donated Big Boy. I wrote and made a request to keep him and the owner was not reluctant at all to let me have him. Sadly, just a short time after I got him Big Boy was found to have heart worms and was ailing. We kept taking him to the vet but he could not do anything for him. My mother took care of him when I was working, and he got a lot of attention but Big Boy was suffering. The vet suggested we put him down but my mother did not want to see that. Big Boy deserved a much better life after what he went through. In the end it was the worst thing I ever had to do and it still hurts today. As a matter of fact, hardly a day goes by when I don't think about Big Boy. All together, I spent almost three years with him (1943-1945).

Postwar Years

After the war I got married in May 1947 on my wife's 20th birthday and we had five sons together. I initially went back to work at US Steel and liked it but I always had the railroad in mind. I had a relative who worked as a switchman for the Elgin, Joliet and Eastern Railway and my father was a railroad man for the Baltimore & Ohio Railroad (B&O) so I ended up spending thirty-seven years with the Elgin, Joliet and Eastern Railway. It ran from the southern end of Lake Michigan all the way (outer belt) to Waukegan, Illinois, which is just short of the Wisconsin state line. We had all steam engines then and I started out shoveling coal as a fireman. Later, I became a locomotive engineer and controlled the big diesel engines (3,000 horsepower), GE and GM units.

Today, I still have shrapnel in my chest from the wounds I received and they took x-rays to confirm it. I do get compensation for it but I never got a Purple Heart. I made a request to my local Congressman and it was denied. I am still fighting for it today.

One of the last reunions for the surviving USMC K-9 handlers was held on 15 October 2008 at the Doberman Pinscher Club of America's National Dog Show in Topeka, Kansas. This event marked the 100th anniversary of the Doberman Pinscher dog in the United States so there couldn't be a better time to honor the service of these brave veterans and their canine partners. Ed is standing in the back row, second from the right. (Ed Adamski Collection)

Reflections

Back then, the Japanese bombing of Pearl Harbor was the main thought on most of the guys' minds and of course, being young, we wanted to kill as many of them as we could. However, we learned pretty quickly that killing came pretty hard and there were other things to contend with, such as, climate, disease and the miserable conditions. In 1972, I organized our first reunion in Lancaster, PA. It took me almost two years to get our guys together. The day we met in Lancaster, the last Jap surrendered.[12] It was a well-known story.

Sadly, Ed Adamski passed away on 1 November 2018, just days before this book went to press.

Right: Ed at his home during the author's last visit. Even after all these years, his memories of his Marine Corps days and Big Boy are vivid and he proudly will share his stories with people who want to know what really happened during the war. (Ron Werneth)

Below: For the author, one of the most heart-warming events that came about as a result of this book was being able to share some of USMC K-9 handler PFC. Ed Adamski's stories with his entire extended family. In July 2016, this special event took place on his 92nd birthday and included several generations of the Adamski family. During this celebration, the author presented the distinguished veteran with two special mementos: the official USMC Dog Record Book and photograph of his loyal K-9 partner Big Boy. Prior to this, Adamski never saw them before; they were discovered by noted War Dog historian Michael Lemish in the National Archives. (Adamski Family)

[1] The First War Dog Platoon, commanded by 1st Lieutenant Clyde O' Henderson, participated in the invasion of Bougainville which began 1 November 1943. They served with the Second Marine Regiment which was commanded by Colonel Alan Shapley. Throughout the combat in primarily dense jungle, no patrols with dogs were ambushed. Two K-9 handlers died fighting on Bougainville, PFC Russell T. Frederich and PFC William N. Hendrickson. (Cyril J. O'Brien, World War II Marine War Dogs, *Leatherneck*, December 1995, 16-18)

[2] All of the dates for Ed's travel to the Pacific was derived from the documents in his United States Marine Corps Service Jacket.

[3] Elephantiasis is usually found in tropical regions of the world and Africa. Mosquitoes can transmit this disease via parasites when they bite a person. The worms block the lymphatic vessels in the human body. This leads to swelling especially in the lower torso including the legs and male genitalia such as the scrotum. There is also a non-parasitic form of this disease that can be caused by regular contact with

soils, especially red clays. These soils have a high content of alkali metals, such as sodium and potassium. They also are related to volcanic activity.

[4] The background on the USMC landings on Guam was derived from, Cyril J. Obrien, *Liberation: Marines in the Recapture of Guam*, Marines in World War II Commemorative Series, Marine Corps Historical Center, United States Marine Corps, 1994.

[5] Topka engaged a group of Japanese soldiers after his Doberman Lucky alerted. Even though he killed ten enemy soldiers, Topka was wounded. Initially, Lucky growled at the medics that came to treat his handler but then allowed them to take care of him. Nevertheless, once the young Marine died, Lucky would let nobody come near his body. Thankfully, Sergeant Edwards (who had spent time with both Topka and his dog) was able to put a noose on Lucky and get him away from his master's body. After this tragic incident, the Doberman was transferred to another handler but just was never the same. Therefore, Lucky was sent back to the United States and was returned to the family who donated him to the Marine Corps.
Captain William W. Putney D.V.M., U.S.M.C. (ret.), *Always Faithful.*

[6] Ed is referring to the Japanese attack on this field hospital when the patients, doctors and corpsmen literally had to fight for their lives. As a result of this attack, thankfully only one patient was wounded but one medical officer and corpsmen later died of wounds. The casualty list also included one medical officer, one dental officer, one Navy warrant officer, twelve corpsmen and sixteen Marines (from the medical companies) were wounded in this engagement. Cdr R. R. Callaway, (MC), *The Third Medical Battalion in Action: Bougainville and Guam,* Historical Study, MCS SenCourse (1948-49).

[7 & 9] www.ibiblio.org

[8] *IsCom WD, 15-31Aug 44, Enclosures C.* These figures include an estimated 500 Japanese civilians that were on the island during this time period.

[10] This background information was derived both from personal correspondence of Capt. Putney (provided by historian Mike Lemish) and Putney's book *Always Faithful.*

[11] The men later moved this camp to a permanent site at Yona, which was on the other side of the island. Ibid. 189.

[12] Ed is talking is talking about WW II Japanese Army veteran Sergeant Shōichi Yokoi. Amazingly, Yokoi spent twenty-eight years hiding on Guam. This veteran felt that as a Japanese soldier, he could not surrender. Then, in January 1972, Yokoi was discovered in the jungle by two local men. After a small scuffle, he was apprehended then gave up. Upon his return to Japan, he became a hero. Yokoi passed away on 22 September 1997 of a heart attack at the age of eighty-two.

Chapter 4

"We didn't suddenly become savages; you had to learn how to fight."

Captain Charles S. Downey (USN)

Several years ago while attending an aviation event in Chicago; I had a chance meeting with a local World War II Navy aviator named Charles Downey. He is one of a very small handful of healthy Midwest-based veterans who served in the Pacific Theatre as a pilot on board an aircraft carrier. Downey was born on 2 August 1924 and grew up on the east coast. During the war, he flew on seventeen strike missions and survived a lethal kamikaze *attack on his ship, the* Ticonderoga. *This veteran now lives at probably the best spot for a lifetime aviation-minded man, at a home built right on the Poplar Grove Airport. He lives above a hangar area where his 1941 Meyers OTW-145 biplane is parked. The walls in his residence are lined with aircraft photos and artifacts, but one framed document grabbed my attention – a personal letter from former President George H.W. Bush. In the letter, the President congratulated Downey on being the youngest United States Navy pilot in World War II.*[1]

Early Years

My basic interest in aviation came from building model aircraft and living in the shadow of the Sikorsky factory (Stratford, Connecticut). I would watch the big Sikorsky flying boat, the S-42, roaring back and forth. I also read a lot and built model aircraft. In those years, there was an annual, statewide, model competition in Connecticut. I participated in this event and one year I was selected to be the outstanding junior contestant. Igor Sikorsky was one of the officials and said, "Young man, it looks like your career will be in aviation – stay in it." He was one of my icons.

Personally, I could not wait to get started after Japan attacked us and killed over 2,000 men in a sneak attack – Pearl Harbor. This was "The Day that lived in Infamy." That Sunday, I was working at a drugstore (Rocky Hill, Connecticut) and making a banana split. One of the gals there said, "Where is Pearl Harbor?" (By this time) the winds of war had been circulating in my head for over two years. There was already a war going in Europe so you knew that we would eventually get involved. The Battle of Britain captured my imagination with these young men saving their country. I had a feeling that I would become a pilot and thought about joining the Navy. My dad served in the Navy during World War I but he did not want to sign for me. Fast forward to June 1942, the Battle of Midway happened and I was just graduating from high school. I knew that I would be drafted and really wanted to be a mechanic on an aircraft carrier. So, one of my dad's friends suggested that I go down to the post office, where the recruiting was being done, and learn more about the naval aviation cadet program. I subsequently went to New York and was screened for this program. You better screen yourself for something; otherwise, you would be drafted as a foot soldier. I didn't want to do that and made the "grade" so here I am. The Japanese were the enemy and pre-flight training changed us from peace-loving Americans into reluctant warriors. We didn't suddenly become savages; you had to learn how to fight.

*****

In 1943, the young airman completed all his flight training and was commissioned as an ensign on 16 July of that same year, at NAS Pensacola.

Left: A young Charles Downey in his United States Navy dress whites. (Charles Downey Collection)

Above: USS Ticonderoga, *the first aircraft carrier Downey served on during WW II. (USN)*

Bombing Squadron 80

Back in February 1944, I was one of fifty-seven pilots that formed Bombing Squadron 80. You look back and the people came from all walks of life. In general, the Navy was mostly full of men from the reserves (both enlisted and officers). However, always the commanding and executive officers were from the "regular" Navy. As a carrier pilot you're always being refreshed. For example, aviators would use a practice deck and make four landings. The *Charger* (CVE30) was based out of Norfolk and would go up and down the Chesapeake Bay. Anyone who needed qualification landings would get scheduled to do four of them. The group commander would say, "You guys have not caught a wire in a while, you better go out to the *Charger*." I did this while waiting to pick up my new aircraft at the factory.

Next, our pilots went to the factory; got our Helldivers, trained in these aircraft and we brought them over to our ship. Then, our ship went to the war. We had religious services aboard ship and were part of a "structured"

Edward L Anderson commanded Bombing Squadron 80 (VB-80) from January 1944 until being wounded on 21 January 1945, as a result of the kamikaze *attack on* Ticonderoga. *Prior to be assigned to Downey's ship he served on board the USS* Enterprise. *While on this carrier, he participated in the Battle of Midway as a Section Leader in Bombing Squadron 6 (VB-6) and flew three missions during this engagement. During one of these missions, his gunner ARM2c Stuart J. Mason, Jr. was wounded in both his face and legs but still continued to fire his guns against the attacking Japanese aircraft. The duo still managed to land safely on the* Enterprise *(CV-6) after Mason repaired the radio in their SBD Dauntless dive-bomber.*[2] *Anderson was later awarded the Navy Cross for the bravery he demonstrated in battle. Then, from August 1942 until joining Bombing Squadron 80, he was Chief Instructor at NAS Jacksonville and Daytona, Florida, where he was teaching young airmen how to fly dive-bombers. After joining* Ticonderoga, *he flew twenty missions between 25 November 1944 until 4 January 1945 and was awarded a second Distinguished Flying Cross. As a result of being wounded in the* kamikaze *attack, he was transferred out of the unit and Lieutenant Paul F. Kalat took over his duties. All together, Edward L. Anderson spent thirty-two years in the Navy and retired at the rank of captain. He passed away on 10 February 1998 but was posthumously inducted in the Virginia Aviation Hall of Fame on 19 November 2002.*

Captain Edward Anderson. (Charles Downey Collection)

society of sorts. You had leadership, protection and the other basics. But, what about the rest of the world? For example, what is the B-29 or the 8th Air Force? When you were fighting a war, you are kind of being "spoon-fed" the information and not privy to everything else that is happening. However, you did hear some rumors while you were aboard your ship. We heard about the *kamikaze* pilots and thought, why are these young kids going out and purposely killing themselves?

When our carrier *Ticonderoga* was on a "shakedown" cruise after the builder trials, one of the four shafts failed. A ship could not be sent to war like this so we got back home on three propellers. We went back to Norfolk and had a three-week delay. During this time, we did extra training. All of our pilots were able to switch aircraft. The dive-bomber pilots got to fly fighters (Hellcats) and the fighter pilots flew our aircraft. I flew a Hellcat and thought, "God, this aircraft really goes!"

Life on an aircraft carrier

You served your six months or longer at the whim of an admiral. If you were in some extended action at the end of the six months, you would be ordered to stay longer. In my case, I was aboard ship from June 1944 until the end of March 1945. You are not in combat that whole time but you were on the ship. There were training exercises in the combat zone if you did not fly for a couple of weeks. They would tow a spar several hundred feet behind the ship and we would attack it. One time I saw a Hellcat disintegrate and strike the water. It was flown by a tough, aggressive, big fellow named Jackson. He kicked the tail off the aircraft. This was before Grumman installed an aftermarket strip (stabilizer) on the tails of these fighters.

We loved Captain Dixie Kiefer so much we would fly underwater for him. He was such a great skipper and always kept us informed. Captain Kiefer was friendly and affable. He loved his troops and was a "perfect" commanding officer. Our ship, the *Ticonderoga*, was his big command (of his military career). Later on, (as a result of the *kamikaze* attack on 21 January 1945), we all cried when he was taken off the ship with all those wounds. Captain Kiefer nearly got killed in this attack and several other senior officers died on the bridge. The next time I saw him was after the war at a big squadron party held at Naval Air Station, Quonset Point, Rhode Island. At that time he was the commander there and we had not seen him since January when he left the *Ticonderoga*. Our old captain still had his arm in a cast (from injuries which occurred on 21 January). He was so pleased that his boys thought enough of him to come and visit.

*****

REGRETABLY, MUCH BELOVED Captain Dixie Kiefer did not really get to enjoy his postwar life very long. He died on 12 November 1945 in a plane crash that occurred in Beacon, New York. Kiefer was once called the "indestructible man" after suffering numerous wounds in his wartime service but he always managed to make it through.[3] *At the time of his death, he was just forty-nine years old and his final rank was commodore. His final resting place is with so many of our other military heroes at Arlington National Cemetery (Arlington, Virginia).*

Captain Dixie Kiefer. (Charles Downey Collection)

SBD dive-bombers in flight. Note the perforated dive-flaps shown in the foreground of this picture (from another SBD). (USN)

An undated picture of veteran Charles Downey flying a SB2C dive-bomber. (Charles Downey Collection)

Opinions of the SBD Dauntless & Helldiver

I THOUGHT THAT THE Douglas SBD Dauntless dive-bomber flew much better than the Curtiss SB2C Helldiver. With this aircraft, you didn't just fly it, you "wore" it. The SBD became "part of you" just like the Corsair did. As for the Helldiver, you "operated" it. This plane did have more range and could carry more bombs but I thought the Dauntless dove better. Nevertheless, you flew the Helldiver to the best of your ability because this is what you had. We did have a civilian tech rep (from the Curtiss Company) aboard to help us. You had a lot of hook bounces with the Helldiver. It bounced on landing and would skip all the wires then crash into the barrier.

I believe there were eight or nine wires on the deck of the carrier. A perfect landing would probably be on the second or third wire. There were four barriers and each one consisted of two wires which could "snarl" an aircraft. The operator of the first barrier was the best and helped save a lot of planes. If he saw a late hook, the barrier would be deployed. We corrected this issue by changing the tail hook controls. Originally, the goofy Helldiver had a complicated tail hook system which was difficult for a pilot to operate. It also had a complicated hydraulic system with four different valves which we called a "plumbers nightmare." This aircraft was called a "Beast" because you had to "tame" it. You had a great job but you had to try your best not to get killed.

A DFC for the young aviator

ON 13 NOVEMBER 1944, American carrier strikes focused on the crowded Manila Harbor, Luzon Island, Philippines. The harbor was filled with numerous ships of the Imperial Japanese Navy including the light cruiser Kiso *along with the destroyers* Kasumi, Hatsuharu, Ushio, Asashimo, Akishimo, Hatsushimo, Akebono *and* Take. *These ships were pounded by over 350 carrier-based strike aircraft from Task Force 38 including Carrier Task Group (TG) 38.1's* Hornet, Monterey *and* Cowpens, *TG 38.3's* Essex, Ticonderoga *and* Langley *along with TG 38.4's* Enterprise *and* San Jacinto. *A morning strike already did a lot of damage to the Japanese ships that were present in the harbor. By the time a second strike arrived over Manila Harbor, as a result of being bombed several times, the destroyer* Hatsuharu *had both fires and flooding, that eventually made the ship settle on the shallow bottom. It was finally abandoned by the crew the following day.* Ushio *suffered a lethal near miss which caused damage, flooding and casualties. The light cruiser* Kiso *also was hit by three bombs in the following places: the bow section, near the boiler rooms and close to the aft gun mounts.*

Lieutenant (junior grade) Downey took part in second strike on Manila Harbor that day. At 1155, Lieutenant Commander Edward L. Anderson and nineteen other VB-80 Helldiver pilots (including Downey) took off from the Ticonderoga. *Flying on the mission with them was a group of fourteen TBM Avengers from VT-80 led by Lieutenant Commander Charles W. Shattuck. Escorting these two groups of Air Group 80 strike aircraft were twelve VF-80 Grumman F6F Hellcat fighters that were commanded by Lieutenant Frank Gooding, USNR. After these three groups of different aircraft rendezvoused, they proceeded together to the target. The escorting fighters flew at 14,000 feet keeping their watchful eyes over the rest of the strike aircraft flying underneath them. The VB-80 dive-bombers were at 13,500 feet while the VT-80 Avengers operated at 13,000 feet. This formation approached Manila Harbor from the northeastern direction and visibility over the target area was reported to be hazy from very thick smoke.*

After arriving over Manila Harbor the aircraft split up and commenced their attacks. The pilots of VB-80 struck four different targets. Downey was part of group of dive-bombers that attacked a light cruiser, described in the after action report as a Natori *class, reported four hits on it. The four VB-80 airmen credited with the "confirmed" hits in the CVG-80s* Aircraft Action Report, 13 November 1944, *included Downey, Lieutenant Commander Edward L. Anderson, Lieutenant (junior grade) George Walsh, USNR and Lieutenant Robert F. Perkins, USNR. Two other Helldiver pilots tried to hit the same enemy ship and it is possible one of them was successful, but this could not be "confirmed." Another two VB-80 aircraft targeted a destroyer that was already damaged on an earlier strike. These airmen reported one direct hit and another near miss. A second Japanese destroyer was the objective of the dive-bombers and four direct hits were observed. This ship burned and sank rapidly. Lastly, Lieutenant George Center, USNR. (of VB-80), strafed and bombed a floating dry dock or repair barge. Flying with him was a veteran newsreel photographer, Norman Alley, who shot 400 feet of footage during the mission. Alley confirmed the sinking of Center's target and described the strike as "beautifully executed." As described in Downey's account one Helldiver and two men were lost due to enemy aircraft-fire.*

As for the fourteen VT-80 Avengers, these planes were armed with 1,000 or 2,000-pound, GP bombs. This group did their attacks using a standard glide bombing technique and released their payloads at 3,500 feet. The enemy anti-aircraft fire below 8000 feet was described as "intense." The land-based targets struck by them included a tank farm, warehouses and a pier. For this strike, VT-80 put in claims for six "serious" hits and another four-six probable hits but it was difficult to confirm the damage due to heavy smoke from the fires below. All of the Air Group 80 strike minus the one lost VB-80 Helldiver returned to the Ticonderoga *by 1540.*[4]

*****

I WAS AWARDED MY Distinguished Flying Cross (DFC) after one of my first missions; it may have been my second or third (in combat). It took place on 13 November 1944. Our target was enemy shipping in Manila Harbor and we were moving up to where it was deadly. There was a big Japanese base in this area and it was pretty well defended. Enemy cruisers were in the harbor and we were assigned to get them. The light cruiser *Kiso* was underway and it was trying to get out, but we got him. I was in the third position for the diving attack on this ship. I saw some of my squadron-mates completely obliterated maybe just 400 feet in front of me. (At that time), I was diving right behind them and the aircrafts flaps were open. All of a sudden, there was a huge flash when their Helldiver received a direct hit from the cruisers anti-aircraft fire and it blew up. The explosion could be measured in nanoseconds and the whole thing just disappeared. There wasn't even any falling wreckage that I could see. (After completing my attack on the ship), my gunner said, "Downey, I think you blew the stack off." At that time we didn't know that the

Kiso was a three-stacker but (after the war), I saw pictures of the sunken ship with two (remaining) stacks and a third one that was knocked off. The bridge also was demolished.

The Imperial Japanese Navy light cruiser Kiso, *in 1942. (S. Kamino Collection)*

In December 1944, Lieutenant (junior grade) Charles Downey received his Distinguished Flying Cross on the deck of the carrier Ticonderoga. *(Charles Downey Collection)*

*****

ACCORDING TO THE METICULOUS research done by Japanese naval historian Tony Tully, the light cruiser Kiso *was hit by three bomb hits earlier in the day of 13 November 1944 before being "shellacked" with over six additional bomb hits in the second strike to attack this group of ships. By 1330, the foredecks of the vessel were awash and it sunk a short time later that day. As a result of the sinking of* Kiso, *eighty-nine men perished and 105 were wounded. However, the majority of the casualties were from the Fifth Fleet staff who was on board this ship, not the* Kiso's *crew. Captain Ryonosuke Imamura also survived the sinking of his ship. This light cruiser was not the only Japanese Imperial Navy vessel to be sunk as a result of Task Force 38's aerial assault on the enemy's ships in Manila Bay. Destroyers* Hatsuharu *and* Okinami *along with seven merchant ships also were lost. Furthermore, destroyers* Akishimo *and* Akebono *were hit while moored at the nearby Cavite Navy Yard. The latter ship was already damaged in a previous air attack nine days earlier and both destroyers ended up sinking the following day.*

Reflections after the 13 November 1944 mission

AFTER THIS MISSION, I thought no more about it but three or four weeks later, I was awarded the DFC. There were two other pilots (Lieutenant Commander Edward L. Anderson and Lieutenant Edward H. Bagley, Jr.) that were also awarded a DFC for this mission.[5] Sadly, some guys never made it and were posthumously awarded DFCs after they perished. The pilot who died (on this mission) was Ensign John "Schoolboy" Manchester.[6] He was a year older than I but had a "baby face" so we called him "schoolboy." As an anecdote, both John "Dugan" Doyle and he completed dive-bombing training about two squadrons (behind our group). They heard that we had not yet gone into combat (on a carrier) so the airmen badgered their training officer to get out of the sequence and be assigned sooner. This is how they joined our group. Later, I made the observation that they both volunteered to join us and got into bad situations. We also lost some aviators in operational training including my roommate. I thought to myself, "God, a guy could get killed out here." I believe in God; however, you were always concerned that you might get killed, but it probably won't happen to you.

VB- 80 Helldiver pilot Ensign John "Schoolboy" Manchester, forever young. On 13 November 1944, he was lost in action over Manila Harbor with radioman/gunner ARM2c John E. Griffith. (Emily Manchester Townes Collection)

JOHN "SCHOOLBOY" MANCHESTER was born on 14 August 1923 at Vanderbilt Campus (Nashville, Tennessee). He had an older brother, Tom, and younger sister, Emily. His father was a teacher at Vanderbilt University and his mother was a housewife. Growing up John wasn't the "scholarly type"; instead he had a love of the arts including playing music, still photography, doing eight-millimeter movies and his own artwork.

The war changed everything for this young man and the Manchester family. While he was a sophomore going to Vanderbilt University, John volunteered to become a naval aviator and enlisted on 1 December 1942. His primary flight training was done in Columbia, South Carolina and he was proud to be the first one in his group to solo on 9 February 1943. After completing this initial training on 14 March, the blossoming aviator went on to the next phase of the curriculum which was pre-flight school in Chapel Hill, North Carolina. This training lasted from 23 April until 26 July 1943. During this time he not only studied to become an airman; it also included a lot of physical activities such as climbing a thirty-five foot rope in thirteen seconds and boxing. After this phase of training, he did another three months of additional work in Peru, Indiana. John then continued his flight training for the next two months in Pensacola, Florida where he practiced essential things to fly as a naval aviator such as instrument flying.

As of the 5 January 1944, he was just twenty-years old but the arduous training was paying off. John now began his final three months required for him to get his "Wings of Gold." This last phase of flight training covered important things such as gunnery, night-flying and dive-bombing. At the end of March, this young man achieved his goal: to get his wings and be commissioned as an ensign but sadly, he would not live to see 1945. From April until mid-August 1944, John shuffled around from Miami, Florida to Cape May and Wildwood, New Jersey. Then, his final stateside stop was Norfolk, Virginia, where he boarded the Ticonderoga *with his good buddy John "Dugan" Doyle and saw his family one last time.*

*****

As mentioned earlier, John Manchester had the nickname of "Schoolboy" given to him because of his "boyish" appearance. During November 1944, his flying partner was radioman/gunner ARM2c John E. Griffith. As mentioned in Downey's account, these men perished together on the 13th of that month when their Helldiver was obliterated by enemy anti-aircraft fire during their mission over Manila Harbor. According to Ensign Raby, their aircraft was hit at 9,000 feet just as they entered their dive on the target. The doomed dive-bomber burst into flames, went into a spin and no parachutes were visible after it was hit. Initially, the Manchester family didn't have much information about what happened to their son John until receiving a heart-felt letter from his buddy John "Dugan" Doyle that gave them more details about his last mission. Posthumously, John Manchester was awarded the Purple Heart but this decoration gave little solace to his family and squadron-mates who remember the promising young man and flying partner John Griffith with a warm smile even after the passing of almost seventy years.[7]

Air Group 80 bags the *Kumano*

On 25 November 1944, Lieutenant (junior grade) Downey took part in a second strike for Task Group 38.3. Their objective was a damaged heavy cruiser in the northern part of Santa Cruz Bay, off Luzon Island, Philippines. The enemy ship was later identified as the Kumano. *Initially, both the carriers* Ticonderoga *and* Essex *were going to participate in this mission; however, the latter ship could not do so because of a* kamikaze *attack just prior to take off. Therefore, only* Ticonderoga's *Air Group 80 provided strike aircraft which consisted of ten F6F-5 Hellcats (eight escorts and two radio relay planes), thirteen SB2C-3 Helldivers and nine TBM-3 Avengers. The overall commander of the strike and the escort fighters was Lieutenant Commander Albert Vorse while the dive-bombers were led by Lieutenant Commander Edward L. Anderson and the leader of the nine torpedo bombers was Lieutenant Commander Charles W. Shattuck. Take off was at 1250 but when Lieutenant Commander Vorse launched,* Ticonderoga *was doing a high speed turn. Because of this, the pilot could not get enough speed to stay in the air so he had to do a water landing but was rescued by a destroyer. Out of the remaining seven Hellcat fighters just three of them along with one photo aircraft and its escort actually "completed" the mission. This is because two of the VF-80 F6Fs had to engage an enemy fighter just after take off. During the ensuing chase by this aircraft, both Hellcat pilots jettisoned their belly tanks and bombs while trying to escape. After successfully fleeing from their opponent, they were not able to join up with the rest of their group. As a result of this, they returned to Task Group 38.3 and joined the combat air patrol over the ships.*

After all three groups of different Air Group 80 (dive-bombers, fighters and torpedo bombers) rendezvoused together, they flew at 12,000 feet as a group towards Cape San Ildefonso. Then, while flying in a northeasterly direction, the strike aircraft split into two formations. The first group consisted of thirteen dive-bombers and two escorting Hellcat fighters that made their approach to the target area while flying in a circling counter-clockwise route. The second group was the nine TBM Avenger torpedo bombers supported by one VF-80 fighter. These ten aircraft broke off in a starboard direction then flew in a wide clockwise direction so that they could approach the heavy cruiser from the north.

*****

When Air Group 80 sank the heavy cruiser *Kumano*, it did a "classic" roll over and then sank. First, our dive-bombers attacked this ship then our torpedo bombers came in. This is what we trained for. We didn't need any fighter cover because there weren't any opposing fighters, but there was anti-aircraft fire. Thirteen dive-bombers attacked the *Kumano* but the last one which was piloted by Ensign John "Dugan" Doyle got hit by enemy fire then landed in the water, maybe 4-5 miles away. He escaped into the jungle on Luzon and was later awarded the Navy Cross for this action. Doyle survived but his gunner; ARM3c William "Rafe" King later died when two aircraft collided. We dove on this ship but I don't know if I got a hit or not. I then watched our torpedo bombers attack. The ship was not underway so it was kind of an "easy" target but they were shooting back fiercely. I could see a torpedo bomber drop his torpedo and saw it hit the ship. After this, the *Kumano* rolled over and sank. When it rolled over you could tell the bottom was so sleek with no drag, like a coke bottle affect. Afterwards, we strafed the enemy survivors (in the water). I had no compunction about doing this; the enemy was the enemy.

Rare photo, taken on board a submarine after these United States Navy pilots were rescued off Luzon Island, Philippines. Left to right: Ensign Maurice L. Naylon Jr. (AG-31), Ensign John "Dugan" Doyle, Ensign Nicholas J. Roccaforte (from VT-18 who went MIA on 29 October 1944 and was declared KIA), Air Crewman P.R. Schleltchy and ARM3c William "Rafe" King. (John Doyle Collection)

ARM3c William King survived being shot down and spending several days in the jungle only to be needlessly killed on 15 January 1945 in a tragic accident. King was flying as a passenger on board a R4D-5 transport which was loaded with medical evacuees that took off from Tacloban Air Field, Philippines. Ensigns Maurice L. Naylon Jr and Nicholas J. Roccaforte also who were rescued with King in this crash. It occurred because an Army pilot was trying to show off his "flying skills" to a nurse in a Piper liaison aircraft when he collided with this transport plane. Everyone on board the R4D aircraft perished as a result of the Army aviators careless flying that day. According to Doyle, the only reason why he didn't fly and lose his life with his fellow rescued aviators was because of a bout with malaria – this twist of fate saved his life. Out of this picture, all of the men except Ensign Doyle and Air Crewman P.R. Schleltchy were lost because of this senseless accident.

Japanese Navy heavy cruisers Kumano *(foreground),* Maya *and* Chōkai, *1939. (T. Kamino Collection)*

Above: 25 November 1944, a view of the destruction of the Kumano *as seen from one of VB-80's Helldivers. (George Walsh Collection)*

Right: Another view of stricken Kumano *as taken from* Ticonderoga's *dive-bombers, just before the torpedoes struck the ship's port side. (Anthony Tully Collection, S.E. Morison files, NH).*

*****

As Downey recalled, *his group of dive-bombers from VB-80 did "very well" during this strike and claimed four direct hits with 1,000-pound bombs. These hits were confirmed by photographs taken of the stricken enemy vessel. The VB-80 pilots credited with successful hits were Lieutenant George Center, USNR, Ensign Joseph E. Kluzinski, Lieutenant Martin H. Kollmorgan, USNR and Ensign John Doyle. An observation noted by some of the VB-80 aviators was that* Kumano *was missing about 30 feet of its original bow from previous attacks and retrofitted with a temporary replacement section. VF-80s Lieutenant Meloy who was flying his Hellcat armed with one 500-pound bomb put in a claim for a hit on the stern of the ship.*

The airmen from VT-80 had a very successful day and put in claims for a total of six torpedo hits which were all done on the doomed heavy cruisers port side. Since these torpedoes were evenly spaced across the vessels hull, it caused massive flooding and damage that caused the ship to roll over and sink just a few minutes later. Lieutenant Commander Charles W. Shattuck who led the flight of nine torpedo bombers and credited with a successful direct hit was awarded a Navy Cross for his actions on this mission. The Kumano *was manned by a crew of 1,036 officers and men. Four hundred and forty-five men perished in the sinking of this vessel including Captain Soichirō Hitomi who chose to go down with his ship.*[8]

A Harrowing mission over French Indochina

On 12 January 1945, the United States Navy sank or destroyed numerous Japanese ships in the South China Sea. On a river near Saigon, we found five enemy ships. Our job was to destroy shipping; land targets came later. I believe we had nine Helldivers on this strike and it took place off Cape St. Jacques, French Indochina. By this time we were already down to fifteen total aircraft and some of them were probably out of commission. Typically, out of nine aircraft, you should get at least four hits. My section leader hit a ship and it blew up in my face at 800 feet. What was I doing flying at 800 feet? I wanted to make sure that I hit this destroyer escort ship. Therefore, I went beyond the 1,500 feet release point and by that time his bomb had hit this vessel and it blew up. I could see a couple of enemy tankers which were still burning from a previous strike. On the way back, I felt very alone. In my mind, it seemed like the longest flight that I ever made but it was only 165 nautical miles.

My aircraft was leaking oil and I didn't know how much was lost because when you are out of oil, the engine simply quits. You had an oil pressure gauge but nothing showed the amount of oil left. So, this became one of my biggest concerns. When would my engine quit? When would I be landing in the water? We did not want to get anywhere near those burning ships because there were Japanese in the water. I still felt reasonably secure because I had a two-man raft (stored between our seats). My gunner's job was to get the raft out if we made a water landing. I did not think we would jump out of the aircraft; that would be crazy. Stick with your aircraft even if it quit, but this did not happen.

After arriving back to our fleet, I could see that the deck was spotted aft and all of our guys had already returned. If the engine quit now, they wouldn't lose us (aviators), but our aircraft would be lost. I thought we could just land in the water by a destroyer and get picked up. But, I wanted to save our aircraft. We were still flying but losing oil. They said, "Stand by" and during this time I was circling around and smoking. This was over the *Essex*, our sister ship. However, I thought that they would have to turn into the wind to get me. They had a Prep Charlie signal which meant get down here and be ready right away. Well, there is no way this big ship was going to turn 80 degrees and get where it should be quickly. It would take them at least two or three minutes to do this. But, you had to get down there and be ready to make the cut. If they were not ready for you, you would have to go around again. Sure enough, they were not ready; this is what I anticipated. It would have been better for me to stay up there. Okay, I thought, now the engine is going to quit. So I eased up the throttle just a little, put the gear up but still kept the flaps down. I went around again and this time, I got the signal to land then caught the #2 wire and shut down the aircraft. Later on, I was handed some fragments that were lodged in the fin area of the engine cylinders. I saved the largest piece which was part of ship's deck house or something. These fragments had cut through some pushrod housings so oil was seeping out and hitting the exhaust system. I asked the Chief how much time I may have had before this engine would have quit. He replied, "Well, lieutenant, we think between nine-eleven minutes.

I was told to report to the Captain. He said, "Lieutenant, you are not quite in the uniform of the day to go in the ward room so you have to stay in ready room." I thought to myself, you bastard, I am out here fighting this war, but saluted and said, "Yes, sir." My enlisted crewman was treated like a king down in the enlisted quarters. He had good chow but I had to eat sandwiches for two days. But I thought I can't make a fuss and slept in a reclining chair in the ready room.

I guess, they were still adjusting to these Marines aboard who were tearing up their aircraft. I also heard two squadron commanders (VMF-213 and another one) arguing about who shot down this four engine aircraft. They thought that it was an enemy *Emily* (flying boat) but it turns out that the aircraft was one of our B-24s. There was bad communication and obviously, bad recognition. This occurred over Indochina and I confirmed the story later after reading about it.

Kamikaze attack

During this stage in the war, Japan was getting desperate trying to fight off the inevitable allied onslaught that was heading towards the outlying Japanese-held territories and the mainland itself. Therefore, drastic measures were taken to try to combat the seemingly endless flow of men and machines which were bringing the war closer to home. The airborne suicide attacks done by both, Japanese Army and Navy airmen with aircraft typically equipped with bombs. They hoped to destroy some of allied armadas of ships, especially the lethal aircraft carriers. These aviators were collectively known as the kamikaze *or "Divine Wind." The background and more details of this program will be explained later in this work.*

21 January 1945 was a bad day for both the young Helldiver pilot Downey and all of the men serving on the Ticonderoga *while they were in the waters east of Formosa (now called Taiwan). As a part of Task Force 38, the carrier was doing air operations against Formosa and supporting the allied landings on Luzon. On that day, visibility was unlimited and this aided the enemy suicide attackers to find their targets. At 1210, General Quarters was announced and* Ticonderoga's *batteries starting to open fire on an enemy aircraft which was coming in. However, this plane got through the anti-aircraft fire and crashed through the flight deck. It was believed to be carrying a single 250-kilogram bomb that exploded between the hangar deck and gallery deck. This caused a powerful fire to erupt in the hangar deck which was filled with parked aircraft and the blaze spread to wardroom passage and staterooms. The hangar deck sprinklers and water curtains helped contain the fire but they could dose the belly tanks which were mounted on the aircraft and started to explode. Luckily, the conflagration started to weaken as it traveled aft and was put out by a water curtain. The brave plane handlers and firefighters worked together to push the burned out aircraft over the side of the ship.* Ticonderoga's *galley, captain's and admiral's store and the admiral's pantry were destroyed at this time.*

After the initial kamikaze *struck his ship, Captain Dixie Kiefer leaped into action and, at 1220, he ordered a course change from 330 degrees to 290 degrees. This course change prevented the wind from making the fire worsen and kept the smoke from going into the fireroom and engine supply ventilation intakes. Kiefer also ordered magazines and other compartments to be flooded. The flooding helped get the fire under control and prevent additional explosions on* Ticonderoga.

Then, at 1258, another kamikaze *aircraft came screaming down on the ship's starboard side from an estimated 2000 feet altitude. Unfortunately, this enemy plane got through the hail of anti-aircraft fire and then passed through the sky forward and sky lookout platform. After going through the lookout position, the aircraft exploded and spewed out fragments all over the forward side of the stack and supporting structure along with the flight deck. Lethal fragments even pierced the gallery deck compartments and deck itself. Furthermore, the number one 5" anti-aircraft director was just about severed at the base and if it fell over, more damage would occur. However, the damaged anti-aircraft director was temporarily resecured with springlay wire cable until proper repairs could be made. Additional fires also broke out throughout the forward part of the island structure and between parked aircraft on the port side of the deck, but later were under control.*[9]

The hole around Ticonderoga's *number one elevator just after the initial* kamikaze *attack on 21 January 1945. Note all the wood splinters on the deck. (USN)*

The Ticonderoga *engulfed in smoke after two successful* kamikaze *strikes on 21 January 1945, but the carrier would live to fight another day. (USN)*

*****

We went out for strikes in the morning and I was getting ready for my afternoon mission. I stood up in my chair because we heard some noise – the 40s. Somebody must have been in close and (by that time), we were already briefed. The "rule" was that you didn't leave the ready room unless you wore your GI hard hat which was hanging above your chair. I was standing up and reaching for it then vroom! We heard a belated General Quarters warning but now I was suddenly on the deck. The impact of the explosion had knocked our chairs out of their holders (near the bases). The guys were trying to figure out what happened. Somebody said, "I think we have been hit." Suddenly, black smoke starts pouring in the ready room and we had no way to shut it off because if you were in the right battle condition somebody elsewhere in the ship has shut off the air system. However, the fact of the matter was that we were caught with our pants down. A few guys were alert and some of them were firing away but we didn't even hear any 20s. In the meantime, a *Judy* dive-bomber armed with a bomb hit the flight deck. It was a classic attack as their instructions were to put a hole in the carrier's deck so we could not launch anymore aircraft, He hit abreast of the island going forward; otherwise, I would not be here talking with you. This *kamikaze* struck the carrier maybe 120 feet from where we were and hit a passageway by the admiral's pantry because there was nothing slowing him down. He exploded in the hangar deck. We had our gas masks on because there was fire and smoke everywhere. At this time, there may have been twelve or thirteen pilots in the ready room. In the end, the senior officer said, "Guys, we don't want to die in this ready room, we got to get out of here. Hold on to the guy ahead of you." I ended up going aft and escaping then went up to our aircraft.

On deck, the gunnery officer said, "Let's get the 250-pound bombs off (the aircraft)." But, we had no luck so I went to a 20mm gun tub on the starboard side with several other Marines. I went behind a protective splinter screen when a Marine sergeant yelled, "God dammit lieutenant, do you want to lose your fuckin' head?" This was because I was bareheaded and not wearing a helmet. In the meantime, a second *kamikaze* aircraft (a *Zero* fighter), about thirty minutes after the first one hit, was fooling around at about 1,500 feet deciding what he wanted to do. I watched him come all the way in as the screening ships were leading the target. That is the reason why there was so much shrapnel on the inboard side of the stack. We had a lot of screening ships protecting us so we hardly fired our own weapons and were in the center position (of this protective group).

*****

21 January 1945 was a day that any veterans who served on the Ticonderoga *would never forget. As a result of the* kamikaze *attack, 144 men perished and over 200 others including Captain Kiefer were wounded.*[10] *The severely damaged aircraft carrier retreated back to Ulithi Atoll for repairs and a respite for the crew.*

*****

In retrospect, the *kamikaze* pilots desire to die for the Emperor was very noble but (quite often) they did not know what they were doing. Many of them could have done a much better job. For example, if they did a 15-18 degree dive, the aircraft would bounce off the (target) ship. There would be a lot of *kamikaze* aircraft that would get shot down within 500 feet of a ship. However, if they came in at 45 degrees or more, the vessel would be hit and there would be a lot of damage. In short, they were a "serious" threat.

As a related sidenote, Downey remembered another incident that happened the same day (21 January 1945) on the carrier Hancock *(CV-19) with a stricken TBM-3 Avenger torpedo bomber. It occurred when the ship was off Formosa. This aircraft was flown by his buddy Lieutenant (junior grade) Carroll "Charlie" Dean along with ARM3c Frederick J. Blake and AOM2c Edward E. Zima. Before departing for the Pacific frontlines, Dean and Downey used to*

The carnage on the deck of the carrier Hancock, *the same day, 21 January 1945, as the* kamikaze *attack on* Ticonderoga. *However, this was an accident on the* Hancock, *which was caused by probably two 500-pound bombs which were hung up on Lieutenant (junior grade) Carroll "Charlie" Dean's TBM-3 Avenger. (NA)*

Burial at sea for forty-three men and seven officers who were lost in this sad accident on Hancock. *(NA)*

drink together at NAS Oceana. This group was about a month ahead of Downey's unit on the Ticonderoga. *On this date, at 1328, Dean's Avenger landed on the* Hancock *and taxied normally on the flight deck. However, as it went by the island, the aircraft blew up. The cause of this explosion was believed to be two 500-pound bombs which were "hung" in the bomb bay then detonated when the TBM landed. The explosion ripped open the flight deck and made a 10' x 16' hole. It also caused damage to the gallery deck area and the inboard side signal bridge. Fires started on the flight, gallery and hangar decks along with three aircraft being destroyed. Remarkably, by 1510 that same day, the* Hancock *had completed the emergency repairs and the fires were extinguished. Forty-three men and seven officers including Downey's buddy along with his crew that manned their torpedo bomber perished in this lethal accident. Seventy-five others were injured as well. Two days later, the men who died were buried at sea. On 25 January, the* Hancock *limped back to Ulithi Atoll for repairs and care for the injured crewmembers.*[11]

USS *Hancock* days

AFTER *TICONDEROGA* WAS badly damaged (from the *kamikaze* attack), we went on the *Hancock* because its air group had a lot of casualties (from combat). So, the survivors (from this latter air group) went home on the *Ticonderoga*. A typical operational tour was six months, give or take a week. (While serving on the *Hancock*), Air Group 7 had a very difficult time with operational accidents. For example, when a fighter plane was getting its wings stowed by two or three men, several .50 caliber machine guns went off and killed five dive-bomber pilots in the ready room. You can imagine the grief aboard this ship so it was a happy time for the guys going home on the wounded

Above: The carrier Hancock *sailing off the Philippines as seen in the winter of 1944 just prior to the time that Downey and his counterparts from the damaged* Ticonderoga *went on board to finish their overseas tour. (NA)*

Right: Lieutenant Downey (left) with his usual VB-80 flying partner/gunner Larry George. (Charles Downey Collection)

Ticonderoga.

We did the first carrier raids on the mainland of Japan. We were frightened because our aircraft would be attacking their homeland. It was February so it was dark, gloomy and cold. We had to wear our scarves and leather jackets. I remember seeing Fuji-san (Mt. Fuji) while flying at 14,000 feet. I was flying with my regular section leader; we always flew together. Our target was Kizarazu Air Base. I said, "Well, if we (dive-bomber pilots) are as good as we think we are, let's destroy it!" We had some torpedo bombers and fighters with us. On that first day, the fighters had a ball and shot down over seventy aircraft. I said, "Why do we need our dive-flaps going down to blow up some hangars? Why don't we get in and get out?" So, we came in smoking downhill without flaps. You felt some comfort not loitering and using your dive-flaps while you were flying at a much higher speed. (On this type of attack), you were diving maybe at 260 knots as apposed to maybe 310 knots. Later, after the war was over, I received a package in the mail with an Air Medal citation for having destroyed a hangar on that strike.

Return to the mainland – spring 1945

We were relieved from the beginning of the Okinawa Champaign and we arrived in San Francisco on Easter Sunday. We were flying the "new" Helldiver models – the SB2C-4Es. Our air group also had Corsairs, -4 models. We were stationed at Chincoteague NAS, where the famous Chincoteague ponies are. We thought that we would certainly be killed out in combat again. Second tour pilots never made it through. So we all had a very fatalistic attitude. Some of the guys got married but I didn't even have a girlfriend.

(For our next tour), we were supposed to go on a modern, large, aircraft carrier, the USS *Franklin D. Roosevelt* (CVB-42), that had an armored flight deck so the *kamikaze* would bounce off. However, the war was over before we could do that. During the summer we did some more flight training. It was around this time that a guy flying a B-25 was lost around Newark and the weather was really bad. This B-25 pilot got lost and then crashed into one of the very top floors of the Empire State Building, which was the tallest building in the world at that time. I was on a training flight around the Chesapeake Bay area. We were not supposed to be listening to commercial radio but who was going to find out? There was a news bulletin that said a B-25 crashed into the Empire State Building and several people were believed to be killed.[12]

When President Roosevelt died in April, Truman became the new President. Then, the decision was made to drop the atomic bomb and after a second atomic bomb was dropped the war was over while we were still at Chincoteague. There was a point system which was based on your time in service, decorations, etc. I was eligible for early release, in September and the G.I. Bill was set up. So all of a sudden, your focus goes back to being a civilian again. You literally just dropped everything and walked away, this included our aircraft and everything else.

Postwar

After the war, Downey went to Rensselaer Polytechnic Institute (Troy, New York) and graduated in 1951 with a degree in engineering. That same year, he got married to his girlfriend Lorna and together they raised five children. She has been deceased for several years but he still has close relationships with his children and extended family members. This veteran aviator never really stopped loving flying or aviation and his choices during the postwar years reflect this. Therefore, he stayed in the Navy Reserve until 1975. Being in the reserves allowed him the opportunity to continue flying a wide variety of military aircraft (from prop-driven ones to jets) until his retirement in 1975 with the rank of captain. For his civilian job, Downey worked for sixteen years as an industrial engineer with American

Airlines. After this, he served as vice president of Commuter Airlines (in Chicago) then worked for Butler Aviation. Lastly, he helped set up Midway Airlines, which was also in Chicago.

As you may expect, Charles Downey, even well in his 80s, continued to fly regularly and loved doing it. Quite often he could be found in the cockpit of his beloved 1941 Meyers OTW-145 biplane or volunteering with the Commemorative Air Force, which is the world's largest flying museum of vintage WW II aircraft.

Chuck Downey and his 1941 Meyers OTW-145 biplane during one of the author's visits. Even well into his early 90s, the veteran aviator was enjoying life to its fullest; just an amazing man. (Ron Werneth)

Reflections

DURING THE 1960S I visited Japan and stayed in the "new" Imperial Hotel. I knew that my bags were looked through from day to day because when I looked for something to write, it would be in a different place (than I left it). The Japanese were still suspicious of Americans especially those in business executive positions. I did not dare wear my Distinguished Flying Cross especially when calling on Japanese aviation officials. When I toured some of their historic shrines I thought of the solidarity of the national spirit of this country despite the fact that it was vanquished. I felt that the old militaristic rule of Japan was now gone.

I've never withdrawn any information that I thought would be helpful in "understanding" the war because I hate it. War is just grizzly and mean; there is nothing "glorious" about it. You have to maintain strength in order just to survive. True Americans are willing to take risks to preserve what we have. However, some people don't realize what it would be like to not have freedom. (During World War II), I did what was expected of me at the time. Nevertheless, when you came home after the war was over, you did not sit around and expect a handout for the rest of your life. We couldn't wait to get home but we knew that the enemy had to be slayed first.

Charles Downey passed away on 19 February 2016.

[1] Initially, it was believed that President H.W. Bush was the youngest Navy pilot in the war because he earned his Navy wings five days before his 19th birthday. However, Downey actually earned his wings seventeen days before his 19th birthday, making him the "youngest" Navy pilot of World War II.

[2] *Bombing Six Action Report, 4-6 June 1942.*

[3] www.arlingtoncemetery.net

[4] All of the specific details for this mission were culled from the CVG-80s *Aircraft Action Report, 13 November 1944.* Historian Anthony Tully graciously shared these resources with me. He also provided the Japanese side of the mission which was put together utilizing a number of official Japanese records and secondary sources.

[5] Lieutenant Edward H. Bagley did not survive the war. He was lost in action on 7 January 1945 with his radioman/gunner ARM3c Joseph T. McHugh. They were flying in SB2C-3, BuNo. 19505 over Formosa. Downey recalled that Lieutenant Bagley was a division leader and a fine senior officer.

[6] Manchester's radioman/gunner was ARM2c John E. Griffith. Manchester and his radioman perished on 13 November 1944 in SB2C-3 BuNo.18698 due to anti-aircraft fire. *United States Navy Combat Casualty Lists of World War II*, kindly provided by historian Jim Sawruk.

[7] A wonderful compendium of John Manchester's letters to his family and life story was put together by his sister Emily. The work included copies of his letters, many of which had detailed accounts of his time in the Navy and showed his personality as well. While going through the book, I noticed a lot of John's doodling of pictures such as aircraft and himself as an aviator. I felt there were a lot of unrealized dreams when he was lost in combat. Emily Manchester Townes, *School Boy to Helldiver: His Life and Letters.*

[8] The background and results of the 25 November 1944 strike on *Kumano* were researched using a number of resources including the careful studies done by Anthony Tully, original Japanese records and CVG-80s *Aircraft Action Report* for this strike.

[9] The background and details of the *kamikaze* attack were derived from the *Battle Damage report of the USS* Ticonderoga, *27 January 1945.*

[10] *Ibid.* Same source as endnote one provided by Jim Sawruk.

[11] The background of the incident that occurred on 21 January 1945 on board the *Hancock* was derived from the *Decklog of the USS* Hancock, *21-25 January 1945.* The casualties were confirmed by Jim Sawruk utilizing the same source in the previous endnote.

[12] On 28 July 1945, a B-25 bomber crashed into the Empire State Building, which was the tallest building in the world at that time. Lieutenant Colonel William Franklin Smith was flying this aircraft on a transport mission when he got lost due to poor visibility and slammed into the building. Smith along with two others on the B-25 perished in the crash. Another eleven people in the building also were lost as a result of this terrible accident.

Chapter 5

"I do not recall having animosity toward German or Japanese people."

Lieutenant Commander George J. Walsh (USN)

George J. Walsh was born in Brooklyn, New York, in 1921. Prior to joining the Navy, he attended parochial schools and Brooklyn College. The distinguished veteran shared some of thoughts during this time period and why he wanted to become a naval aviator.

*****

The war in Europe contributed to my restlessness. I became convinced that we would be drawn into war. We had uncles who had been in the First World War and I had been exposed to volumes of picture books which depicted the horrors of trench warfare. I wanted no part of that fighting. My romantic summer novels tilted me toward the mounted knights in shining armor or the flying aces of the sky dueling in the clouds high above the muck. In 1940 long before the draft I began applying to the Navy for training as a naval aviator. They rejected me initially because I was only 128 pounds, under the minimum weight for any height. Persistence paid off and I was accepted pending completion of college.

*****

In 1941 George enlisted in the Naval Aviation Cadet program and was called to active duty immediately after Pearl Harbor. After flight training at New Orleans and Pensacola in 1942 Ensign Walsh was commissioned a naval aviator. Remarkably, he learned how to fly even before getting his driver's license. George graduated number one in his class at Pensacola with 4.0 scores on his class work and no down checks from his flight instructors. Then, he volunteered to be a pilot in the Marine Corps but two other "sturdier" men were picked instead of him. The veteran feels this may have saved his life because he may have died flying combat missions from Guadalcanal.

The young airman's first assignment after graduating was with VS-34, which flew anti-submarine patrols from Floyd Bennett Field in Brooklyn. Besides doing these patrols off the east coast, the unit had to escort convoys, merchant ships, oil tankers and troop ships. This duty was important because of the never-ending threat from the undersea German U-boats. The pilots of VS-34 flew OS2U Kingfishers that had wheels

Lefdt: Navy dive-bomber George Walsh in flying gear. (George Walsh Collection)

Below: George (on the wing) with his VB-80 flying partner/gunner ARM 1c Virgil Gordon. (George Walsh Collection)

fitted, instead of floats. George was a member of this unit until being reassigned to VB-80 in late 1943. According to George, the initial VB-80 roster of 24 January 1944 listed fifty-seven pilots, sixty-eight enlisted personnel, and administrative officers Bill Clayburgh, Wiley Moore and Ken Price. However, out of the original fifty-seven pilots that helped form the unit only seventeen pilots served the squadron's entire tour of duty.

Throughout his time in the Pacific with VB-80, ARM 1c Virgil Gordon flew as his gunner. George recalls a quiet, soft-spoken New Englander from Guilford, Maine. Because of his short stature, he could be overlooked in a crowd but was totally dependable as a shipmate. However, Gordon's "true" character came out later in the war when the ship held boxing matches. He was fierce in the ring because the young man had been a Gold Gloves Champ!

Combat with Bombing Squadron 80

From the time the squadron formed in early 1944 until leaving for the Pacific Theatre on 29 August (of that same year), Walsh's experiences were similar to the other VB-80 veterans in this section so it isn't necessary to reiterate this background. However, I really wanted to know more about the combat mission that he flew on 13 November 1944 and later was awarded the Distinguished Flying Cross for his actions that day. This is another perspective of the mission flown by fellow Helldiver pilots Charles Downey and George Center.

*****

On 13 November 1944 Fast Carrier Task Force 38 of the Third Fleet steamed 160 miles east of Manila in the Philippines. Its mission was to provide protection of MacArthur's landing force on Leyte by neutralizing the airfields and ports of Luzon. Our group was assigned the center of the island, including Clark and Nichols Fields, the port of Manila, and Corregidor.

We launched before dawn and each plane rose to slide into squadron formation by the light of a rose colored rising sun, which became visible over the horizon as we gained altitude. We throttled back to a slow climbing speed to conserve fuel and gain altitude. Flying west toward Manila we had to reach 14,000 feet flying over the snow-capped mountains of eastern Luzon. Our flight included twenty-four dive-bombers, two divisions of twelve each. The divisions included sections of three planes in "V" formation, and I led the last section of three planes. We were loaded with 1,000 pounds of bombs.

After we crossed the mountains and the plains of central Luzon we increased speed, dropping down to about 12,000 feet as the city of Manila came into view through scattered white cumulus clouds. A heavy layer of smog and smoke covered the city of Manila and the harbor. The skipper, Lieutenant Commander Anderson, turned toward the southern end of the city and sent our division, headed by Lieutenant Paul Kalat, toward the northern end. As we began a high speed run in at about 12,000 feet nothing was visible but a pall of overcast smoke. Kalat signaled to me for a let down through the overcast for a glide bombing run from lower altitude and began his descent.

Just as I was about to follow with my two wing men, I spotted a doughnut hole of an opening in the clouds to the west and headed toward the clearing. Suddenly, I had an open view to the sea below. Two miles below me a Japanese warship was steaming at high speed toward the open sea. It was centered in the doughnut hole of the clouds as if sighted on a target range. Without a second thought I signaled attack to my wingmen, peeled off, dropped my dive brakes, adjusted the throttle, blower and propeller pitch, opened the bomb bay doors, armed the bomb, and pointed the nose of the plane straight down at the target, lined up stern to bow.

At 300 knots thirty seconds of the two mile dive passed in what seemed to be slow motion speed as the black puffs of exploding anti-aircraft shells floated by, punctuated by the red tracer bullets of machine gun fire. The dive brakes hold the speed of the plane from approaching high velocity as it would in a free fall or power dive. The pilot is pressed forward against his shoulder straps because the aircraft is held back as if suspended from a rope. There is time to adjust the aiming point by using the elevators and ailerons as the ship grows bigger and bigger in the windscreen. That day there was no wind factor to be compensated.

A cruiser is a narrow target and I stayed in my dive until I was confident of scoring a hit, and released the bomb. At that speed another two seconds would have made me a *kamikaze* suicide pilot. I pulled out hard; probably 13Gs, low over the water, and taking evasive action while I retracted the dive brakes, adjusted the throttle, blower and pitch, closed the bomb bay, and raced south toward the rendezvous. ARM 1c Virgil Gordon reported a direct hit but I did not look back; I often wonder why. I guess my instinct was to get the hell out of there and back to the protection of the group. We were also so low over the water all my attention was occupied in flying the plane, looking where I was headed and watching out for other possible planes in the area, including Japanese fighters.

I flew southeast toward the city. Gordon, my gunner, was equipped with a K-10 camera. As photographic officer, my plane was fitted with a K-56 camera in the belly of the fuselage. I flipped the plane up on a wing, pointing the bottom of the plane at the city, and clicked off the shutter a few times on the way south to the rendezvous. After the flight back over the mountains to the carrier, Ensign Ben Case landed first because his gunner, ARM 1c Jorgensen, was wounded. We had a quick lunch and twenty bombers were launched again for a second strike.

My Distinguished Flying Cross citation reads in part, "Leading his section in a bombing strike against enemy shipping, Lt. Walsh pressed home a low altitude attack through intense antiaircraft fire to score a direct hit on an enemy cruiser which was left burning and sinking."

A view over Manila Harbor during the morning of 13 November 1944. (George Walsh Collection)

Another perspective over Manila Harbor on the same date but in the afternoon. (George Walsh Collection)

A Photo Mission Experience

GEORGE SHARED ANOTHER notable mission from this same time period that was not an "attack" on an enemy shipping or land-installation. Instead, he had to fly a low-level and slow, photo mission. On 11 November 1944, Air Group 80 participated in an attack on a Japanese convoy attempting to reinforce Leyte after General MacArthur's initial landings in the Philippines. The entire convoy, consisting of four merchant ships carrying troops and ammunition, and four destroyers, was sunk in Ormac Bay west of Leyte. By 14 November, the Army had established a beachhead on Leyte and was looking to the next phase, the invasion of Luzon. They needed photos of the city of Manila but the sky was continually cloudy and high level photography useless. VB-80 was assigned the task of taking low-level photos of the city of Manila.

*****

I HAD AN SB2C EQUIPPED with a belly camera, and a red-headed photographer mate named Davis was assigned to replace Virgil Gordon in the rear cockpit. Our instructions were to make mapping runs over the city at 2000 feet. Two fighters were to be our escort, while the rest of the air group would make a diversionary attack on Cavite Naval Base. The air group had attacked the Manila waterfront and shipping in Manila Bay for several days, meeting intense anti-aircraft fire.

The group took off and headed east over the mountains. A heavy cloud bank covered the area. As we approached Manila at 14,000 feet I broke off into a shallow dive into the clouds while the group continued southeast to attack Cavite. When I broke out of the overcast directly over the city there was no sign of the fighter escort. We were alone and proceeded to make the straight and level mapping runs north and south at 2000 feet.

On the first passes there was little opposition. I guess the defenders were holding fire from the anti-aircraft batteries, because we could see the anti-aircraft and bomb explosions across the bay at Cavite and they would not waste heavy ammunition on a single plane. We could clearly see the Japs on the rooftops firing at us with machine guns. Then, heading north the third pass, we began to get the puffs of heavy caliber anti-aircraft fire exploding ahead of us. We were flying at a slow cruising speed and they were leading us too much. Then the concussions began to rock the aircraft and I turned violently to the right. The straight and level mapping orders were automatically cancelled as my reaction was as instinctive as ducking a punch.

We were weaving as we turned south again, and this saved us; not only from the anti-aircraft threat. In a banking turn to the west Davis called on the intercom, "Mr. Walsh, there's an *Oscar* (The Allied code name for a Nakajima Ki-43 fighter) on our tail." The *Oscar* had been making a textbook attack on us, moving up directly below and behind in our blind spot. The turn uncovered him. Pushing the throttle to the wall we raced up into the clouds and lost him by staying in the clouds all the way to the mountains where we rejoined the group headed for the carrier.

The bombing squadron Daily Action Report stated that many valuable low altitude photos were obtained. The fighting squadron's Action Report also stated that one section of two fighters followed the photo plane as escort but we never saw them. They certainly never came below the cloud layer. Other fighters covering the bombers at 10,000 feet reported that one silver *Oscar* made a pass at them and disappeared in the clouds. At the 1989 *Ticonderoga* Reunion (in Charlestown, South Carolina), Howard Chamblin turned up for the first time. He was the ship's photographic officer and reported that this mission created a discipline problem for him. Davis announced that he would never accept another photo assignment like this even if it meant a court martial.

*****

George and his flying partner Virgil Gordon continued to fly combat missions together until the lethal kamikaze *attack which occurred on 21 January 1945. After this, they transferred to the* Hancock *and resumed flying duties again. By February 1945, Lieutenant Walsh was a very experienced Navy dive-bomber pilot and one of his significant wartime experiences was taking part in the Battle of Iwo Jima in the skies above the island.*

Mission over Iwo Jima

We struck Iwo Jima for two days prior to the invasion. It was like flying over a pock marked moonscape. By this time our planes were equipped with rockets mounted under the wings. With a normal bomb load and our twenty millimeter cannon we were formidable for attacking ground targets in support of troops. The problem at Iwo was that we could not find any targets. The day before the Marines landed I flew from one end of the island to the other at two hundred feet looking for something to attack. Finally, the controller told me just to unload all my bombs and rockets on a sector marked on my grid map of the island.

The day of the invasion I was tasked to take a movie camera man in my rear cockpit to film the action. Our guns were removed and climbed in with an arsenal of motion picture cameras. We had an assigned altitude because the ships off shore were shelling the beaches and inland. We flew back and forth over the landings as if we were in a stadium. The guns of the ships blasted the shore. Fighter planes zoomed in to release torrents of flaming napalm. It was impossible for anything to be alive under all that punishment.

The troops disembarked from the landing ships and were loaded into the landing craft. Then the landing craft formed circles. The circles then separated into straight lines and the rows of landing ships moved onto the beaches. They hit the beach exactly on time. H-hour looked so easy from us flying above that we departed thinking the invasion was a snap. It was not until the next day that I began to read of the true sacrifices of the Marines on Iwo Jima.

One of VB-80's Helldivers flying from the carrier Hancock *on 19 February 1945 during the invasion of Iwo Jima; note the two battleships in the water below. (Charles Downey Collection).*

*****

SHORTLY AFTER THIS, on 1 March 1945, Bombing Squadron 80 flew its last combat mission of the war. Their target was enemy shipping at Amami ŌShima, Nansei Shotō Isles (Ryūkyū Islands, located southwest, between Kyūshū and Taiwan). By this time, the unit had operated continuously for fourteen months. After this mission, the men boarded a jeep carrier then went to Guam, where they spent a few days. Then, they boarded the British carrier HMS Ranee *and went to Hawaii. Next, from Hawaii, the group boarded a passenger liner and traveled to San Diego, California.*

George's last Navy assignment (while on active duty) was being stationed at the Naval Carrier Air Acceptance Test Center (Mustin Field in Philadelphia, Pennsylvania). The aviators were supposed to test the arresting gear of new aircraft before they were accepted by the Navy for flying duties. However, according to the veteran, in reality, this was more of a utility squadron. Therefore, he was able o fly a nice variety of aircraft from a Piper Cub to a twin-engine Cessna. It was the most fun that he ever had flying in the Navy.

Once the war finally ended in early fall 1945, George applied for discharge from active duty but remained a member of the Navy Reserve. In recognition of his service in Navy, George J. Walsh was awarded the Distinguished Flying Cross and several Air Medals. In 1958, he retired from the Navy Reserve with the rank of lieutenant commander. During his flying career, he flew twenty different types of the Navy's aircraft, including single and twin-engine, land and seaplanes, including the Stearman Yellow Perils, OS2U Kingfisher, SBD Dauntless, Curtiss Helldiver SB2C, Grumman Hellcat F6F and Grumman Widgeon.

Postwar

As a civilian, George founded and operated an advertising agency in New York City. He also worked as investment advisor but now is fully retired. George and his late wife Anne had three daughters and one son together. They also have five grandsons and one granddaughter.

Reflections

Looking back to the early days of World War II, we were all concerned about the threats posed by German, Italian and Japanese aggression. The accounts of Japanese atrocities as they invaded China and German disregard for civilian casualties in Europe horrified Americans, even before word of the holocaust began to leak out. Pearl Harbor and the relentless drive of the Japanese in Southeast Asia, the Bataan Death March, as well as the German bombing of Britain bred a fear among us that our own world might soon be in danger.

I do not recall having animosity toward German or Japanese people. We feared the power wielded by Hitler, Mussolini and Tojo. It was a "force" we had to fight to defend ourselves; it was impersonal. As we dove our aircraft into the anti-aircraft fire of Japanese warships we were not targeting gunners. At the same time the Japanese blazing away at us were not trying to kill a pilot but to destroy the bomb carrying airplane that threatened them. Was it really any different in ground warfare with tanks and artillery?

*****

This is a song that was written about the Curtiss Helldiver aircraft when Bombing Squadron 80 first started flying the aircraft. You can easily understand their feelings about it during this time period, just prior to flying the dive-bomber is combat. Even after all these years, VB-80 pilot George Center recalls it very vividly.

Oh Mother, dear Mother, take down that Blue Star;
Replace it with one that is Gold.
Your son is a Helldiver driver,
He'll never see thirty years old!

Now the men who are working for Curtiss
are frequently seen good and drunk;
One day with a hell of a hangover,
they mustered and designed The Clunk.

Yes, the wings were built for precision,
and the fuselage so strong it won't fail.
But, who were the half-witted morons
who designed the cockpit and tail?

Our Skipper hates Helldiver drivers -
he doesn't think much of the Clunk;
Each time that we fly off his carrier,
he prays that his ship won't get sunk!

High, slow, fast, low -
"Wave him off!" he screams to the LSO.
Low, slow, fast, high,
God, won't they ever learn how to fly?

Chorus:
Bring back, bring back my body to me (repeat)
Now my body lies under the ocean, now my body lies under sea.
My body lies under the ocean, wrapped up in an SBC.

George Center was born on 16 September 1921. While attending Georgia Institute of Technology (Georgia Tech) from 1938-1940 he took Civilian Pilot Training (CPT), both primary and secondary courses and this caused his grades to suffer. On 20 September 1941, George enlisted in the Navy Reserve V-12 program which had an age requirement of twenty years old, but it took him three attempts before making the minimum weight requirement of 133 pounds! While awaiting flight training, he was assigned to NAS Atlanta where the novice airman taught other cadets in the fundamentals of navigation.

Next, from November 1941 until being commissioned on 18 June 1942, Center was stationed at NAS Corpus Christi, Texas. After this, the young Navy officer married his sweetheart Eleanor Stark Hosch who he met while being a member of the Kappa Sigma fraternity during his college days. Remarkably, their wedding took place in a Methodist chapel, conducted by a Presbyterian Minister and Eleanor was a Baptist so you can say it "covered all the religious requirements."

The young couple bought a used 1937 Lincoln Zephyr sedan and went back to NAS Corpus Christi where George served as primary flight instructor flying N3Ns and Stearman N2Ss for a few months. Then, he transferred to VS-504 (later to become VS-35) in Bermuda and his unit flew OS2U Kingfishers (on floats) doing anti-submarine patrols. Next, in 1943, his group transferred to NAS Norfolk where they did test flying of the Curtiss SO3C Seamew on wheels. Sadly, their skipper, Commander Seaborn was killed during a test flight and this helped persuade the cancellation of the under-powered aircraft. After this, the airmen flew the SBD Dauntless and transferred to Beaufort, South Carolina where they flew anti-submarine patrols.

In late 1943, Center was assigned to Bombing Squadron 80 (VB-80), which was being formed in Wildwood, New Jersey. His background during this period is similar to the other airmen from his group who are profiled in this section. VB-80 arrived on board the Ticonderoga *off the Philippines in early November 1944 and started their flying operations in the Pacific. George was considered to be an excellent airman by his peers and he was a division officer in the group. During the late fall of this year, he had several notable experiences in combat. This aviator took part in the 13 November attack of enemy shipping and land-based facilities at Manila Harbor. While on this mission, Lieutenant Center had Norman Alley, an accomplished newsreel photographer, as a passenger on his aircraft. Alley described the shots as the "most dramatic scenes of destruction" that he ever saw and felt the attack was beautifully executed. He also confirmed that Center had sunk a floating dry dock or barge after strafing and bombing it. Another important mission for VB-80 occurred later that same month on 25 November which resulted in the sinking of the heavy cruiser* Kumano. *Lieutenant Center's Helldiver was the fourth aircraft in the dive-bombing attack and he was credited with a direct hit. After the enemy ship rolled over, George saw its (now exposed) four propellers. For his action on this day, the airman was awarded the Distinguished Flying Cross.*

By early 1945, George Center and had racked up plenty of experience flying missions as a dive-bomber pilot. One of his biggest honors as a naval aviator was the Silver Star which was awarded to him because of his actions during the mission on 12 January 1945 strike off of French Indochina (as mentioned in Downey's story). During this strike, Lieutenant Center led a coordinated attack on Japanese shipping while his group of aircraft was being hit by heavy anti-aircraft fire. According to his citation, he scored a direct hit on a destroyer which caused it to explode and disintegrate.

After World War II ended, this veteran Navy pilot stayed in the service flying everything from jets to many different multi-engine aircraft before retiring in May 1964 as a commander. George and his wife Eleanor enjoyed sixty-nine wonderful years of marriage until her passing on 18 November 2011. Together, they had two sons and one daughter. In retrospect of his years as an airman, George said, "I was pretty lucky. Since the age of Lindbergh's flight, I always wanted to fly. It is also important to remember the gunners who flew with us; they rarely even faced forward (while in combat). Jim Booker (now deceased) was my gunner and he was a superb individual."

The death of a ship. VB-80 in action off French Indochina. (George Center Collection)

Chapter 6

"Even today, I can remember the face of the young pilot that I shot down."

Lieutenant Kunio Iwashita (IJN)

Back in the year 2000, one of the first World War II Japanese Navy fighter pilots that I met was Kunio Iwashita. My first impressions of him were that he was very intelligent, soft-spoken and kind. At that time, Iwashita-san was attending a symposium being held at The National Museum of the Pacific War (Fredericksburg, Texas). During his visit with fellow Japanese veterans Jirō Yoshida and Taisuke Maruyama, I did my best to assist them. The event itself was great and everything went well for these men during their stay. For me, this would be the "serious" beginning of over fifteen years of work to try to document the lives of these often misunderstood or forgotten veterans of a war fought so many years ago.

After I decided to move to Japan, Iwashita-san became one of my great supporters, and while serving as the president for The Zero Fighter Pilots Association he introduced me to many of his distinguished counterparts. One of things the former naval aviator told me over the years that always stuck into my head was this, "Whenever I flew combat missions, I always thought this could be my last day alive. It always was scary."

It was during my time in Japan that this remarkable gentlemen gave me his personal diary and said, "Please use this work instead of trying to interview me. I wrote my experiences down in great detail and this is the best way to tell my story." I was very honored to do this; but regretably, Iwashita-san passed away on 8 June 2013 at the age of 92 years old before he could see this in print. I want to take this opportunity to say thank you, old friend. I apologize

*Veteran Japanese Navy fighter pilot, Kunio Iwashita.
(Kunio Iwashita Collection)*

for not getting this book done sooner. The story below is edited from his personal diary and numerous meetings/ correspondence with him over the years. Much of the difficult Japanese terminology and background research was done with the help of Minoru Kamada and Ikuo Shimizu in Japan. Unfortunately, most of the original flight records held at The National Institute for Defense Studies in Tokyo, Japan that covered this veteran's combat missions (except for Iwo Jima) were found to be missing.

Early Years

I WAS BORN ON 1 March 1921 in Iida City, Nagano Prefecture. Iida City is located in the central part of the main island of Japan (Honshū), in a high basin surrounded by mountains that were 3,000 meters (9,843 feet) high. I find my life a bit strange having been born in a mountainous area and never seeing the ocean until I was twelve years old. However, I later joined the Navy then became a pilot. (At that time), my family owned a restaurant. It is still in business, run by my youngest sister, and the restaurant will soon be 100 years old.

I was born as the fourth, out of six children in our family. I had three brothers and two sisters. I was small in size, but nimble. I entered middle school with the highest grades possible and was the class president every year. Since my brother (Yutaka) entered the difficult-to-get- into Naval Academy, people said, "The Iwashita brothers are smart." When I was a child, the average living standards in Japan were much lower than they are today. It was very difficult for parents to have their children get an education beyond middle school. Therefore, the military academies which were supported by the government were very popular. The Navy would accept anyone with an excellent record, without any discrimination of their family or past relationships. That was the reason why the Navy had such excellent personnel. I think it was one of the best things about this branch of the military.

On 3 November 1937, I received a telegram that I was accepted into the Naval Academy. My classmates, Mitomō Marutō and Masatoshi Yoshioka, also were accepted. Remarkably, only ten middle school students (including us) from the entire Nagano Prefecture were admitted to the Academy: three from Matsumoto Middle School and one each from Nagano, Suzaka, Iiyama and Suwa Middle Schools. In the past, only one student from Iida Middle School was admitted per year. These former alumni included: Shōichi Hitoharu (62nd Class), Saburō Nishina (64th Class), my brother Yutaka (66th Class), Kaoru Kumagaya (67th Class) and Mikihitō Nakagawa (his former name was Yoshikō Harashi, 68th Class). So, it was very unusual for three of us to be admitted into the Naval Academy in one year. Because of this, Iida Middle School became the top school in Nagano Prefecture.

During this time (in Japan), militarism was at a very high level. Young students would admire the Navy officers with their sharp uniforms and belt-mounted daggers when they came home for their yearly vacations. After seeing

September 1938, at the Naval Academy, the Iwashita brothers, Kunio (right) as a first year student while Yutaka (left) in his last (fourth) year. (Kunio Iwashita Collection)

them, I vowed to work harder so I could be admitted to the Naval Academy as well. When I received my entrance letter to attend there, I felt like my spirit soared to the heavens with pride. Among my buddies, we felt that the best schools in Japan were the First High School (in Tokyo) or the Naval Academy but the latter was more difficult to get into.

According to the official records, the number of young men accepted into the Naval Academy were the following: the total amount applicants were 7,134 young men; out of this number, 3,764 candidates passed the physical tests while only 1,230 applicants were successful on their written tests. In the end, only 354 people were accepted into the Academy that year. This number shows that only five percent of the total applicants were admitted to my class, the 69th. Furthermore, the applicants were limited to students who were not only physically qualified, but also academically in the top tier of their school. If you compare the above numbers to other years, it was 3.5% for the 65th Class, 3.3% for the 66th Class, 3.5% for the 67th Class, 4.0% for the 68th Class and 6.3% for the 70th Class.

The Naval Academy was located on Eta Jima, a small island off the naval port of Kure. At the center of the island is Mt. Furutaka which faces Kure Bay. All of the students lived in dormitories and were secluded from society. Everything we did, day and night, was focused on becoming an officer in the Navy. For three or four years (depending on the class), candidates lived in a very strict, spartan environment that was supposed to fill them with the spirit required to be an officer. During their time at the Academy, upperclassmen led their junior counterparts, while the latter group tried to support each other.

My class graduated six months earlier than usual because of the war and what was happening around the world at that time. After three years of training, on 25 March 1941, 343 trainees completed their classes at the Academy and were classified as candidates to become future ensigns. On graduation day, we immediately boarded a training ship and sailed to such places as Amoy (now renamed Xiamen, located on the southeast coast of the Chinese mainland) and the Palau Islands, then returned to our base at Yokosuka. Because of World War II in Europe and the unstable situation in the Far East, our training cruise was shortened to only six weeks. Only the battleship *Yamashirō*, along the cruisers *Nachi*, *Haguro* and *Kitakami* took part in this cruise and it was the last one in Navy history.

Iwashita-san was on the Haguro *during this training cruise.*

After sailing into Yokosuka, all of us went to the palace and paid our respects to the Emperor. Then, we were assigned to a respective ship. My assignment was to be a navigation officer on the heavy cruiser *Aoba*. At that time, Yasuhirō Nakasone (later prime minister of Japan) was the officer in charge of the supply department. He was a lieutenant (junior grade) when we served together. Nakasone was well-known as a respected and highly-opinionated officer. We did intense training to learn the proper techniques for doing torpedo attacks and I became a real "officer" in the Navy.

Reflections on the Imperial Japanese Navy

Since its beginning, the Imperial Japanese Navy was modeled after the British Royal Navy. Because of this, our Navy used a lot of English words in daily life. Sailors who couldn't speak a word of this language used English words on a regular basis such as wash tub and gun room. In the Academy, we were taught proper table manners for official dinners, complete with silver tableware so we wouldn't appear impolite abroad. There were many senior-level students from my school that were assigned to be pilots. Some of them went to the United Kingdom for a few years of flight training then stopped by the United States for observation, on the way back to Japan. My father-in-law, Rear Admiral Katsuji Kondō, was one of these veterans.[1] Although, beginning in 1930, he studied in England for four years; Kondō-san seemed to believe it was absurd for a small country like Japan to fight against the United States since he knew the power of America. One night, when I was training to be a fighter pilot in Ōita, Kyūshū, he spoke to me in his private room. At that time, he was a captain, and said, "There will be great difficulties ahead in this war." There were many people like him in the Navy who were knowledgeable on foreign affairs and they tended to be pacifists (such as Admiral Isoroku Yamamoto). Nevertheless, I'm afraid that these sensible thinking men were not listened to; so, in the end, the Army and Navy hardliners started this war very carelessly The end result was the loss of a lot of wonderful and highly-skilled people.

Pilot Training Days

On 1 November 1941, I was promoted to the rank of ensign while still serving on the *Aoba*. Next, I was ordered to become a student pilot with the Kasumigaura *Kōkūtai*. Our class was the 37th Term and it consisted of 137 officers who all graduated together. We studied the basics of flying and flew a training aircraft (Navy Type 93) nicknamed "The Red Dragonfly." While I was still a senior at the Academy, I took aptitude tests for becoming a pilot and did some flight training at Iwakuni Air Base. I believe this is the reason I was eventually chosen to become an aviator. During my training, my brother Yutaka was one of my flight instructors. He was a senior when I entered the Academy. After Yutaka did his training cruise, he was assigned to be a member of the 34th Term Class at Usami Air Base. Because he was one of the best student pilots, Yutaka received a ceremonial silver watch from the Imperial Family. After he completed his flight training, Yutaka was assigned to be an instructor. Because of this, he trained me. My brother was a very tough instructor and many of my colleagues recall being poked by his baton or scolded, via the gosport communication tube, when they flew with him.

After learning the basics of flying, the date for my first solo fight was posted. The important things were the taking off and then landing properly (doing a three-point style landing at the right altitude). I was told this many times by my flight instructors during my training. For beginners, it was difficult because in the middle of descending at 15 meters (49 feet) per second you had to quickly determine the proper altitude to lift the nose (for a three-point landing). If you lost control, not only would you destroy a precious aircraft, you also would hurt yourself. We trained ourselves by checking the land formations while going to and from Tokyo. For example, "It is xx meters away; that place is xx meters away." I was very nervous when I soloed for the first time. I felt relieved and content after my successful takeoff and landing at the airport.

In the middle of night on 7 December, Yutaka wanted to meet me at the Kasumigaura Shrine. There, he was waiting by himself and very solemnly said, "Tomorrow morning we are finally going to war with the United States, beginning with an attack on Pearl Harbor." I almost couldn't speak after I heard this news. On 25 July 1942, my brother was assigned to be a *buntaichō* on the carrier *Zuikaku*.[2] When he left for the frontlines Yutaka left me locks of his hair to remember him by, because he might not return alive. (While on the carrier), he participated in several battles including the Battle of Santa Cruz. During this engagement, he attacked the enemy carrier *Hornet* and his aircraft was hit by enemy fire. Sadly, his observer was killed while Yutaka was seriously wounded. Regardless of his injuries my brother was able to push himself and reach friendly airspace. Once he reached there, Yutaka lost his strength and landed in the water near the destroyer *Yugiri*. A rescue crew from the ship picked him up but sadly, he died soon after that on 26 October 1942. I heard about my brother's last moments from Inoue-san, a 68th Term (Naval Academy graduate) and a navigator on *Yugiri*.[3]

At the end of August, I completed my flight training with the Kasumigaura *Kōkūtai* and was assigned to be a fighter pilot with the Ōita *Kōkūtai* with thirty-five other aviators who completed the training with me. The rest of the trainees from my original group become aviators or air crew members for either bomber or seaplane aircraft and subsequently transferred to either the Fussa or Hakata *Kōkūtais*. At Ōita, we did very intense training including gunnery, night-flying and practiced doing formation flying. During this training, we flew the Type 96 fighter with fixed landing gear (Allied code name *Claude*). We learned the skills required to be a fighter pilot in combat. It was around this time that I began to think dying in combat would be inevitable for me. This thought really hit me when, on 18 November, out of the blue, Lieutenant Sadamu Takahashi, (*kanbaku* commander from *Zuikaku*), flew to our air base to personally tell me about my brother's death in combat. After that, I trained even harder and expected that one day I also would die in combat.

Left: Kunio's older brother Yutaka Ishimaru (he changed his family name after getting married) was killed in combat on 26 October 1942 during the Battle of Santa Cruz. (Kunio Iwashita Collection)

Below: 27 February 1943, Graduation Photo of the Flying School Students, at Ōita Kōkūtai.
Front row, left to right: Lieutenant Teiichi Mukai, the fifth person in the row is Captain Katsuharu Kondō, next is Commander of the 12th Combined Naval Kōkūtai, *Rear Admiral Sadaichi Matsunaga, Commander Chujirō Nakano, Lieutenant Commander Kaneo Kunisada, Lieutenant Commander Minoru Suzuki* (Taicho). *Second row, left to right:, the second man is Lieutenant (junior grade) Takashi Oshibuchi, then, Lieutenant (junior grade) Kunio Kanzaki and Lieutenant (junior grade) Iwashita. Third row, left to right: the second person is Lieutenant (junior grade) Yoshishige Hayashi. Lastly, the middle officer in the fourth row is Lieutenant (junior grade) Yoshio Ōba. As Iwashita-san stated, most of these men died during the war and he was just one of a handful of survivors. (Kunio Iwashita Collection)*

In 1943, while serving as a flight instructor of the Ōita Kōkūtai, *Iwashita-san (right) together with Lieutenant (junior grade) Motonari Kaneko. (Kunio Iwashita Collection)*

On 1 November 1942, we were all promoted to the rank of lieutenant (junior grade). Then, on 27 February of the following year, we completed our flight training without any accidents. To my surprise, I was awarded the honorary silver watch from the Imperial Family which was presented to me by Prince Kuni Asaakira. I was proud of my skills as a fighter pilot but my gunnery skill wasn't that good so I didn't expect to get this award. Of course, I was happy to receive it but really thought this award was given to me because of my brother's death. However, I wasn't very happy about my next assignment, to be a flight instructor with Ōita *Kōkūtai* for the 38th Term Class. Out of the 137-person class, only thirty-five men survived the war. As for the fighter pilots, out of thirty-five aviators only four did not perish in combat. Even though these airmen were the "best of the best" at twenty-three or twenty-four years old, they would ultimately die during the war. Finally, on 5 November, after almost nine months of being an instructor, I received a new assignment. While looking at this telegram, I jumped up and down; my new assignment was to become a *buntaichō* with the 301 *Kōkūtai*. To this day, I still have this telegram in my desk drawer.

301 *Kōkūtai* Days

On 12 November, after one year and three months of being a pilot trainee and flight instructor, I joined the 301 *Kōkūtai*, which was stationed at Yokosuka Air Base. Even though I only served as a flight instructor at Ōita for a short time, a few of my classmates already had died in the war. Most of them lost their lives in air battles in the Solomon Islands. When I was teaching my students how to survive air combat, I also gained experience in that type of fighting. I felt it helped me survive difficult situations in combat (later on). Furthermore, during this time, I had two physical concerns. First, I had lost consciousness a few times while doing high-speed aerobatics. The other issue that I had was a terrible stomach ache so I had stay in bed for a few days. The first problem occurred because my body was subjected to 5gs (while flying at such a high speed and then lost consciousness). Gravity caused a lack of blood circulation to the brain. However, veteran aviators could prevent this from occurring by tightening up their lower body, which stopped the blood from reversing circulation, just before they lost consciousness. If you were a novice pilot and did not know how to do this, you could end up crashing into the ocean and dying. In my case, especially due to the non-stop flight training with my students, I was too physically tired to have enough strength in order to do something about it. Sometimes, I would regain consciousness and realize my flying was erratic so I would try to quickly turn back my aircraft and land as soon as possible. After that experience, I tried to take care of my body and restore my physical strength, hopefully, to prevent this from happening again. The second problem, my stomach ache, was caused by my appendix and I will explain more about that later.

My 301 *Kōkūtai* did not fly *Zero* fighters; instead, we flew the *Raiden*. It was a new shorter-range fighter that was designed by the Navy in order to fight bomber-type aircraft. In 1940, when I was on summer vacation I fell in love with the *Zero* when I had a chance to see one in a hangar at Yokosuka Air Base. However, at that time, I never thought I would get a chance to fly this plane just a few years later during the war. Nevertheless, several of my classmates would later die flying this fighter and I had several near-death experiences with it myself. In the end, I felt my destiny was with the aircraft. This amazing success in combat surprised the world and our enemies began to fear the *Zero*; it became the best fighter (of that period). We were proud to be fighter pilots and had a lot of confidence in this aircraft. Even though the *Zero* achieved a great deal of superiority in combat, the Navy rushed to design a new fighter, the *Raiden*.

After the United States started to deploy heavy bombers flying in large formations, flying at higher than 10,000 meters (32,808 feet); they started to carpet bomb and this caused heavy damage to both our land-based installations and ships. As the days went by, the *Zero* fighters became less effective in combat and could not prevent the enemy's high-altitude bombing attacks. Therefore, there was a need to develop an affective short-range fighter which ultimately led to the development of the *Raiden*. It was designed to have good climbing capabilities and to fly at high altitude. The fighter had heavy fire power required to engage heavy bombers. In order to accommodate these requirements, the *Raiden* was less maneuverable than the *Zero* fighter. Furthermore, while dreaming about dogfighting against enemy Grumman fighters, a lot of our aviators wanted to fly a *Zero* (not a different aircraft) so they were disappointed when seeing this new plane that had a bulbous-shaped fuselage and large engine.

Regardless of our disappointment with the aircraft, we focused on doing training flights with the *Raiden* day and night so we could be proficient with it. We had a very hard time flying this fighter and lost five aviators in six months even before going to the frontlines. For example, a ground crewman was struck by a propeller when a *Raiden* was taking off because the pilot couldn't see him due to a lack of visibility from the front of the aircraft. Furthermore, due to engine problems on a few occasions, these fighters had to do emergency landings and were damaged heavily. Even though we experienced these concerns with the plane, by May of 1944, our group improved our flying skills with it. Therefore, forty-eight *Raidens* were ready to be deployed via Saipan to go to Iwo Jima and the southern theatre.

On 15 March, I was promoted to the rank of lieutenant; then, at the end of the following month, I married Setsukō. She was the oldest daughter of Captain Katsuji Kondō. Since we were at war, I was able to spend just one month with my wife before leaving for Tateyama Air Base with my squadron-mates. We were supposed to stop at Iwo Jima and then go to the Solomon Islands. However, it was in the middle of the rainy season and there was a low pressure system over the region. When we passed Izu Oshima, our cockpit windshields were filled with dark clouds and we could not continue flying. Despite our enthusiasm and will to continue on, we had to fly back several times. With each return to Tateyama, there were a number of accidents because this base had a narrow runway.

While we were waiting for the weather to clear up, on 15 June 1944, the Americans started landing on Saipan; this changed everything. Therefore, we decided to gather our forces at Iwo Jima so we could engage the enemy at Saipan and support our forces there. In order to do this, our Navy consolidated my group, the 301 *Kōkūtai* with the *Hachiman Butai* (or "God of War Unit").[4] To carry out this contingency plan, we gathered all the *Zero* fighters together and gave all our well-maintained *Raiden* fighters to the 302 *Kōkūtai* because the *Zero* was better suited for this mission.

Next, on 21 June, we advanced to Iwo Jima. It was there that I risked my life while engaging some Grumman fighters. Early on the morning of the 24th, about sixty F6Fs from the carriers *Hornet* (CV-12) and *Yorktown* (CV-10) suddenly appeared and started to attack us. The enemy was already flying over our airspace and was in the advantageous position as we were still on the ground. During three scrambles, I lost five of my comrades including Lieutenant Katsumi Kōda and Lieutenant (junior grade) Korekiyō Kawakita.[5] In addition to this, between 3-4 July, more than one hundred aircraft attacked us daily. After receiving a Morse code message, "Enemy Air Raid on Chichi Jima," I took off and climbed up into the dimly light sky. Before reaching 100 meters (328 feet) altitude, enemy fighters from above, started to fire at me. However, I was able to climb high enough and escape them just as it became daylight over Iwo Jima. As I scanned the area below I saw the results of the American attack; smoke was in the sky, and some aircraft and fuel tanks were on fire. As I tried to find some enemy fighters, I flew lower above Mt. Suribachi when I spotted four fighters but they looked like they were Japanese. When I was about 100 meters (328 feet) away from them, I was surprised to see that they were actually star-marked Grumman Hellcats! As I got closer, about 10 meters (32 feet) away from the enemy fighters, they still didn't seem to recognize me so I opened up with all four of machine guns on one of them. As my bullets tore through this fighters wing, the air was filled with flying debris. At that very moment, I saw the face of the enemy pilot as he looked over at me while his Hellcat was engulfed in flames. Then, the stricken fighter went straight down and crashed into the water below. Several Hellcats went after me and I could hear loud noises as bullets tore through my windshield. I could see that there were numerous bullet holes in my left wing so I did an emergency landing at our base on Iwo Jima. Soon after I landed, several mechanics ran to my aircraft and congratulated me. They said, "*Buntaichō*, you did a great job! Those four fighters ruthlessly attacked us just before you came! We shouted *banzai* as you shot down the enemy." This was my first experience in combat and I was lucky. Even today, I can remember the face of the young pilot that I shot down.

By this time, only ten of our pilots including the commander from my unit were still alive and the runway was pock-marked with lots of holes. After the attacks were over, we felt some relief. I could see with my naked eyes there was a huge fleet of battleships and heavy cruisers which were sailing from the east towards a position off the central portion of Iwo Jima. After seeing this, one of my ground crewmen felt some relief and shouted, "*Buntaichō*, we are okay now! Our Combined Fleet is here to help us!" However, as I looked closer, I saw the stars and stripes on the ship flags. Immediately, I shouted loudly to everyone, "Retreat to Mt. Suribachi!" As soon as I uttered this, the enemy fleet started shelling us. The only choice we had was to try to take cover until the shelling stopped. The enemy attack went on for a long time and so intense; really indescribable. Since we expected the Americans to land the next morning, I thought this would be my last day on Iwo Jima. On the contrary, the enemy fleet moved southward and no enemy landings occurred. We cleaned up our air strip including picking up unexploded ordnance so that our planes could land again. After a few days, the ten of us from the 301 *Kōkūtai* returned to Kisarazu Air Base (in Japan). Then, we were promptly transferred to several other air units. Out of the ten survivors, only eight lived to see the end of the war. Nevertheless, the majority of the men in my unit including people who did such duties as maintenance, ordnance, communications, etc. stayed on Iwo Jima. Sadly, by the following March, all of them would die in combat while fighting the enemy.

As an anecdote, the reason why the enemy frequently had the upper hand in combat was the fact that we were way behind in radar technology. We never knew beforehand when they would be coming. As a result, we lost thirty-one fighters, eighty percent of our total aircraft from the 301 *Kōkūtai*, in just a three-day fight.

*****

*O*N *4 J*ULY *1944, there were two missions flown by this unit (*Fighter 601 Hikōtai *of the* 301 Kōkūtai*) that day over Iwo Jima. The eleven* Zero *fighters that participated in the first engagement took off at 0440 just ten minutes after enemy aircraft appeared on Japanese radar. However, one aircraft had to return due to engine trouble. The overall mission was led by veteran fighter pilot Lieutenant Iyozō Fujita. This small group fought against an estimated 160 American aircraft and lost three men that day, WO Iwao Mita, PO2c Takayoshi Morita and CPO Masami Iwatsubo. At 0600, the seven surviving* Zeros *returned to their base. The same Japanese fighter group (*Fighter 601 Hikōtai*) later flew a second mission that same day when at 1330, three* Zeros *took off to engage an American flying boat. Iwashita-san actually mentions this mission at the end of this story in his reflections. According to their flight records, they were successful in shooting down this aircraft.*[6]

*The losses for the American side of this first air battle on 4 July consisted of five aviators. One man was lost due to anti-aircraft and the other four men to enemy aircraft. The losses are the following: from VF-14 (*Wasp*), Ensigns Joseph H. Dougherty (to AA) and Alberto C. Nisi.and from VF-31 (*Cabot*), Lieutenant (junior grade) Haig G. Elezian Jr., Frank Hancock Jr., and Malcolm L. Loomis.*[7] *Much later in life, Iwashita-san was very keen to find out more about the young American aviator that he shot down in this battle and possibly visit the family of the fallen veteran. Several years ago, The United States Navy Historical Society researched this request on his behalf but it was impossible to be certain which person he shot down. Remarkably, Iwashita-san and his family ended up meeting several relatives of Alberto C. Nisi. to pay their respects and share some memories of a fallen pilot from this aerial engagement so many years ago.*

Going to the Philippine Frontlines

AFTER TAKING A TEN-DAY vacation in Japan, I was assigned to be a *buntaichō* in the Fighter 701 *Hikōtai.* The primary duty for our unit was to escort a recently formed T-Attack Force. Lieutenant Commander Hideki Shingō, the *hikōtaichō* of group, was famous for his experience and bravery as a fighter pilot. Furthermore, under my command, there were two men, with the rank of lieutenant (junior grade) and were both 71st Term graduates (from the Naval Academy). They had recently completed the difficult training at Ōita Air Base. The goal of the T-Attack Force, which was comprised of veteran aviators, was to attack the American mobile fleet in the frontlines of Taiwan and the Philippines. We hoped this would change the tide for our losing war.

We were surprised to learn that our aircraft for this unit would not be the *Zero* fighter; instead, it would be the recently deployed *Shiden.* As I described earlier, in 1940, the *Zero* fighter was clearly a superior aircraft in combat. Meanwhile, the United States tried to design aircraft that would be better than Japanese planes. In the beginning of 1944, the enemy sent the new Grumman Hellcat to the frontlines of the Pacific Theater. Once the Hellcat started flying combat missions, the once superior *Zero* fighter started to lose in battle. However, Japan designed a new fighter called the *Shiden*, which was eventually superseded by the *Shiden Kai* (or "improved *Shiden*"). Just as the Hellcats arrived, our *Shidens* were deployed as well. I had the heavy burden of being one of the first ones to use this new fighter in combat. Its Homare (1,825 hp.) engine was especially troublesome. Furthermore, this aircraft had a very complicated landing gear design and, frequently, we had troubles with it during takeoffs and landings.

Hikōtaichō Shingō didn't like the *Shiden* from the very beginning and never wanted to fly it. Nevertheless, one day, out of the blue he said to me, "*Buntaichō*, find me a mechanically sound *Shiden* so I can fly it today." He then proceeded to take off nicely and his test flight went well. However, when Shingō-san came into land, one of the landing gears didn't retract and his fighter was damaged. The following day, he went to the Navy Personnel Office and requested a transfer to a unit that was using *Zero* fighters stating that I should take his place in the Fighter 701 *Hikōtai.* This all occurred "out of the blue" for me and due to the pressing war situation, I didn't have many choices.

As a junior *hikōtaichō*, with the help of some seasoned airmen, I did the best to improve my group. However, after about a month, I had a sharp pain in my stomach. The following day, I had an operation to remove my appendix at Yokosuka Navy Hospital. In addition to all of this, while on his way to Kyūshū, my *buntaichō* almost died due to serious injuries from a landing accident. Nevertheless, Commander Ayao Shirane, a 64th Term Naval Academy graduate, was hastily transferred to lead this *Shiden* unit in the Philippines, due to the importance of their duties there. In September 1944, when typhoons were hitting Japan, he arrived (at our base). Fighter 701 *Hikōtai*, led by its third *hikōtaichō*, advanced to Mabalacat Air Base (on Luzon) in the Philippines. Our unit did combat missions against the enemy forces who were landing at Leyte. Sadly, Commander Shirane died while attacking a United States Navy torpedo boat off Tacloban.[8] After hearing this news, I felt that I should have died instead of him. So, while I was still recovering from my operation, I went to the Navy Personnel Office and made a request to be the next *hikōtaichō* of the Fighter 701 *Hikōtai*, which was granted. In early November, I left Miyazaki Air Base with twelve other *Shiden* fighters and we flew via Okinawa and Taiwan to Mabalacat Air Base. When I arrived, my pilots were very happy to have me as their *hikōtaichō* again.

I would like to talk about the "change" in what was happening in the war. As I mentioned earlier, initially in combat, our *Zero* fighters performed splendidly and were very successful. However, around the end of 1943, such as in air battles during the Solomon Islands Campaign, the results were reversed and the American side began to win. As the war progressed, it was obvious that we were losing a lot of *Zero* fighters in aerial combat. Basically, this occurred because of the industrial might of the United States (compared to our country). A lot of time was required to train our fighter pilots and the lack of fresh replacement men caused our veteran aviators to stay on the frontlines until they died in the war. Furthermore, between one-third, to at worst case, one-half of our pilots did not return from engagements against the superior American fighters. Countless veteran aviators were dying in combat but our Navy had to send less-experienced men to replace them. Therefore, these "green" pilots experienced even worse loses in combat; it became a vicious cycle. The once exceptional reputation of our *Zero* fighters began to fade away.

As for our opponent, the Americans, they had more than enough aviators to deploy fresh and well-rested men to the frontlines. They also were able to get non-combat breaks (or leaves) on the mainland or Hawaii to get sufficient time away from the frontlines. The United States also had the technological advantages in both radar and communications equipment. This enabled them to properly coordinate and utilize large numbers of F6Fs in the air at one time so they would be successful in combat. In April 1943, Admiral Isoroku Yamamato, Commander-in-Chief of Combined Fleet, while doing a tour of the frontlines to encourage our forces, was shot down by P-38s over Buin. Then, in March 1944, his successor, Admiral Mineichi Koga died after his aircraft crashed during a flight from Palau to Davao. The war situation (for our side) became worse and worse as it progressed.

I witnessed the enemy's superiority in the air with my own eyes at Mabalacat (located inside the Clark Field complex, Philippines). A P-38 used to be easy prey for our *Zero* fighters but this time, the P-38s were chasing our aircraft. When I saw this, I was shocked. Furthermore, Lieutenant Yukio Seki, who graduated from the Naval Academy a year after I did, was successful in doing a *kamikaze* attack on the American fleet. This occurred about a half a month before I arrived in the Philippines. It is possible this young officer could have been motivated to become a *kamikaze* pilot because Seki-san realized that we would ultimately lose the war. He could have wanted to do something honorable before we lost.[9]

Around Clark Field (Clark Air Base), there were a lot of anti-Japanese guerilla fighters and native Filipinos that hated us. So, many Japanese aviators who had to do emergency landings outside of our base were robbed and killed. We felt isolated and in danger because there were no safe places outside of our own base. As the days went by, there was less and less food for us to eat. Our main food source was Japanese sweet potatoes. All of these facts indicated that we were losing the war in the Philippines.

Under the previously mentioned conditions, we would fly missions to escort our transport ships going to the Leyte frontlines or engage enemy aircraft over our base. Therefore, we really couldn't achieve much success in combat. I only recall one day in December, when my *Shiden* unit, which was equipped with Type 3 phosphorous bombs, was successful in shooting down three B-24 bombers. It was surreal seeing these large four-engine bombers slowly spiraling down then crashing into the ground below. Nevertheless, during combat in the Philippines, our *Shiden* fighters were sometimes lost due to friendly fire because they looked similar to the enemy Grummans (Hellcats), both in fuselage shape and color.

During this time period, our *Shiden* unit took part in the vital daytime combat missions against the American landing operations at Mindoro Island. My unit, the Fighter 701 *Hikōtai*, subsequently officially became a part of the 341 *Kōkūtai*, due to the losses of both men and aircraft. As the days went by, the war became worse for us. By this time, two of my classmates, Lieutenants Yamada and *kamikaze*-attack leader, Fukabori had died in combat. The latter man perished by crashing his aircraft into the enemy warship at Leyte Gulf. Knowing all of this, I really didn't expect to come back to Japan alive.

After Leyte was conquered the Allies started attacking Mindoro Island, which was midway between Leyte and Manila, and a suitable "stepping stone" in the Philippine invasion. December in the Philippines was more like early fall in Japan. Our humble officer quarters were located on a hill that overlooked the vast plains around Clark Field. We kept the cremated remains of our men who had died in small urns until they could be shipped back to their homes. As I looked at the illuminated houses below, I prayed for my country.

On 15 December 1944, I was assigned to command a group of aircraft including some of our *kanbaku* based in the Clark region. As soon as we finished our breakfast, we took off. Twelve *Shidens* formed up in the sky above but the total amount of aircraft was less than twenty planes. This was the only aircraft we had operable at that time. We flew in a southward direction with the *kanbaku* flying at a height of 6,000 meters (19,685 feet) and our escort fighters were 500 meters (1,640 feet) above them. The sky was clear and we kept our eyes open for both the American landing forces as well as enemy aircraft while flying towards Mindoro Island. Before this, I had never seen an American invasion fleet, but I had imagined it would be in a circular formation with groups of landing boats. Just as I thought it was about time to reach the enemy landing forces, I spotted them. I was shocked to see the countless landing boats with very little space to maneuver in the water below. I figured even if our bombs and bullets would hit every boat, the damage would just be minimal to them. The minute this thought came to my mind, I spotted some reconnaissance P-38 fighters flying under our group. After I dropped my external belly fuel tank, I dove down at full speed towards the enemy fleet while our bombers were gradually descending to begin their attacks. Just as I did this, my aircraft was jolted by the anti-aircraft fire from the enemy ships. As our *kanbaku* descended through this anti-aircraft fire, they attacked the warships, which were doing evasive maneuvers to protect themselves. I felt that our attack was a successful because we caused considerable damage to the enemy while losing only a few of our aircraft.

However, during the mission, we got into thick clouds and I lost our bombers so I panicked. Since I couldn't see any of our bombers, I joined up with the other *Shidens* and decided to go back to base. With our fuel consumption in mind, we flew at low altitude to try to avoid any engagements with the enemy. Just as I was about to fly over some mountains (on Luzon), all of sudden, my engine stopped. Immediately, using my left hand, I switched the fuel lever to the other tank. Fortunately, my engine started running just as I was heading to crash into a mountain-side. Altogether, our losses for this mission were very minimal; we lost only one fighter because of either engine trouble or running out of fuel.

Tragedy in the Philippines

4 January 1945 is a day that I can never forget. Thirteen *Shidens* had just flown in from the Japanese mainland and were lined up on the ground with their engines running. These respective fighters were getting ready to fly an escort mission for an Army transport convoy anchored in the Lingayen Gulf. While these airmen were getting the details of their forthcoming mission from the flight leader, I began to feel uneasy when I spotted two enemy fighters flying in nearby western skies. It almost seemed like they were waiting for just the right moment when all of the pilots would be sitting in their aircraft to attack. These enemy fighters dove down and starting strafing our *Shidens*. As I realized what was happening, I tried to yell as loud as I could for our men to get out of their aircraft but the loud noise of the running engines drowned out my voice. In a split second, after the screaming of the enemy rockets, all of our *Shidens* were engulfed in flames. Many pilots died in their cockpits without any chance to escape while others had serious injuries and tumbled out of the aircraft. As our fighters began to burn, bullets which were stored in the wings, started to explode one after another; it looked like a scene out of hell. Even though it was dangerous, I dashed to the area and was able to pull out two aviators who were still alive. I saw firsthand the power of the enemy attack. A pilot who lost one hand and leg, died in my arms due to a huge loss of blood, uttered, "I want to fight....want to fight." Those two cursed enemy fighters were shot down by ground fire when they tried to attack again from low altitude; I admired their bravery. Our losses that day, both from a physical and psychological standpoint were horrendous. A few days later, the United States started landing operations in the Lingayen Gulf and combat operations in the Philippines entered its final stage.

On 9 January 1945, the United States starting landing operations in the Lingayen Gulf, which is northwest of Clark Field. We said farewell to our last *kamikaze* attack pilots by waving our camps. We thought that the Allies would land at Manila Bay, but they landed in the Lingayen Gulf, which was north of previously mentioned bay. Our headquarters couldn't really determine what the Allied forces were planning to do next. Japanese military forces couldn't even do any sufficient reconnaissance to better understand what was happening. I know this fact is true at least in the case of my unit, the 341 Kōkūtai; we didn't get any information about the American landings (at that time). I believe maybe only the Combined Fleet Headquarters might have known about these landings. However, with that being said, we really didn't get too rattled. Sooner or later, we expected to do hand-to-hand combat against enemy ground troops. Believing that this time had come, about five hundred men of the 341 *Kōkūtai*, led by Commander Tadao Funaki, built fortifications on the hills west of Mabalacat. This was done with the anticipation that enemy forces would attack from the south. The total number of our forces was about 15,000 men so we really didn't think we could win. The name of our group was the KBK *Rikusentai* (Clark Air Base Navy Ground Forces). It consisted of men from eight *kōkūtais* along with personnel from air depots, Navy Stores Department, Navy Civil Engineering Department, Navy Construction Department and the 1st and 2nd Air Flotilla Headquarters. However, despite the large amount of men, we didn't have enough food or weapons; we also never did training to prepare for ground combat.

The Air Chief-of-Staff thought of the idea of sending twenty bags of twenty kilograms of pumpkin seeds from Kisarazu Air Base, which would be dropped by parachute to us. We were able to collect sixteen of these bags and everyone planted these seeds with the hope of growing some pumpkins to eat We also each had one piece of salt. Our group built defense fortifications and we could see the base area from a distance away. Even though we built all of these fortifications, we only had Type 38 rifles and about one hundred grenades for 15,000 soldiers. Compared with the weapons and equipment the Allies used, it looked like we were using our bare hands to fight tanks. With our primitive weapons compared to what the enemy used, it is easy to figure out the eventual outcome. We planted our air-dropped pumpkin seeds behind our defense fortifications and watered them using a nearby stream. Eventually, when these seeds started to sprout, we felt some satisfaction, but this didn't last very long. On 22 February, we were ordered to retreat far behind our initial fortifications so in the end, our efforts to grow our own pumpkins were for nothing.

Without much of a fight from our forces, the Americans occupied Clark Field from the south at Lingayen. Once they took over this base, enemy fighters and mechanized forces could attack us at will. As we gradually lost our positions, we retreated farther into the jungle. Most of the 30,000 Japanese troops lost their lives while retreating from the enemy. Our men suffered from starvation, malaria and dysentery. The leader of our unit, Commander Tadao Funaki was said to have survived until 14 May, but by that time he was all by himself. On 16 September, at the end of the war, thirteen of our soldiers surrendered to native Filipino forces and were treated well. They were transferred to an American POW camp and by late November returned to Japan. This isn't a precise number, but all told, about 800 Army and 450 Navy troops survived in the Philippines and surrendered. But, this was only a mere four percent of our total forces that served there.

The above information was based on the following sources: a diary of a quartermaster from the 341 *Kōkūtai* who returned from the Philippines and a Mr. "Y" who was an Air Staff officer with the 26th Air Flotilla. On 18 June, the latter man with four of his comrades, surrendered to American forces on the beach, in central Luzon. In 1993, he sent me this information about our last stand at Clark Field and I feel that it is very accurate. The other 341st veteran recorded some abhorrent things that our men did to get food that I cannot share with anyone. I heard that today he is living in peace and serving as a monk. As I learned more about the miserable fates of the men who served with our KBK Defense Forces, I realized how futile and meaningless this war was. This is the reason that I wanted these stories to be heard by future generations. After the war, the Filipinos erected a memorial for our *kamikaze* pilots; I am very surprised that they did this.

As I described earlier, our Navy forces at Clark Field were annihilated. My unit, the 341 *Kōkūtai*, hid our men in the defense fortifications on the hill behind the base to be ready for the coming battle. I was ready to fight. However, on 10 January, around midnight, I was summoned by our commander and immediately reported to his office. He told me about an emergency order from the Combined Fleet Commander. This order stated that all available fighter aircraft should be moved to our base at Tuguegarao in order to attack the enemy ships anchored in the Lingayen Gulf. Our chief maintenance man said there were only three or four fighters available for this operation. Initially, I suggested to our most experienced veteran aviator, Lieutenant Commander Iyozō Fujita, a 66th term Naval Academy graduate, that I was the best man to lead this mission. However, Fujita-san strongly opposed my suggestion. He said, "During my combat missions at Pearl Harbor, Midway and the Solomons, I lost many of my pilots, even though I did my best. You are still young and you have to fight for the future of Navy Air Force. I want you to fight so that Japan could change the tide of this war. Don't worry about me, you should go." Finally, our commander decided that I should to go to Tuguegarao; even it was against my wishes.

Without getting any sleep, our aircraft mechanics prepared four *Shidens* to fly. I was full of remorse as I left the base with the other pilots at dawn. After a thirty-minute flight, we all safely landed at Tuguegarao Base. Remarkably, I had been with Commander Fujita since my first combat mission with the 301 *Kōkūtai*. While we were together at Iwo Jima, Fujita-san told me to stick with him because he expected a tough fight in the air. When I was attacked by the Grumman fighters, this aviator saved me and put himself in danger. He really watched out for me. One of the reasons Commander Fujita cared about me was because he graduated the same year at the Naval Academy as my brother. Maybe, he considered me to be like a younger brother? Because of this, Fujita-san saved my life twice.

Lieutenant Commander Iyozō Fujita was truly an amazing and fortunate fighter pilot ace in the Imperial Japanese Navy. Even though he participated in numerous combat missions from the beginning of the Pacific War from Pearl Harbor until Japan's surrender, Fujita-san survived and in his postwar days became a pilot for Japan Airlines. (Iyozō Fujita Collection)

Surprisingly, we saw each other at Echague. The reason why this happened is because, later on, all of our aviators at Clark Field were ordered to go to Tuguegarao. They did this by sneaking through enemy positions. Then, our headquarters wanted to send them back to Japan via Taiwan (at that time called Formosa) so they could be ready for aerial battles over the homeland. Several hundred pilots rode on trucks, avoided attacks from local guerilla fighters by traveling at night and were successful in reaching Tuguegarao in a few days. The 341 *Kōkūtai* including Commander Fujita were part of this group.

After reaching Tuguegarao, most of the above-mentioned aviators returned to Japan within a short period of time. For a while, my *Shiden* unit did some reconnaissance missions to observe weather conditions and enemy troop positions. At that time, the Americans had not attacked Tuguegarao yet, so for a brief time, we had some peace there. We had enough food such as rice, sugar and even some tobacco. So, I could finally eat rice again! I'll never forget how delicious rice tasted after eating only sweet potatoes for such a long time. Tuguegarao was a small peaceful town and we spent a few uneventful days there. We could even go out and mingle with the town residents. This was much different than our time based at Clark Field. Sadly, all of a sudden, American fighter aircraft showed up and destroyed our four beloved *Shidens*, which were carefully hidden in the jungle.

Even though I had lost my *Shiden*, I was able to complete my orders and return to the Tainan *Kōkūtai* by one of our aircraft. Then, at the end of January (1945), I was assigned to be a *buntaichō* in the Yokosuka *Kōkūtai*. Even though my time in the Philippines was less than three months, I took part in difficult combat and experienced death on several occasions. Our pilots always received priority to return back to Japan from the frontlines. While on the other hand, our ground-based troops had to stay and fight without any way of returning home. In the end, these men lost their lives in the war. My unit, the 341 *Kōkūtai* finished up miserably as I mentioned earlier. I am pretty sure that Tuguegarao was attacked again by American forces and guerillas. After I returned to Japan, I thought about the fates of the young medical officers and journalists who were left behind. As I thought about them, my mind was filled with anguish.

Yokosuka *Kōkūtai* Days

During this time, Yokosuka was the most technological-advanced bases in the Imperial Japanese Navy. Our *kōkūtai* was comprised of highly-skilled pilots who were very experienced aviators. Our unit tested new fighters and other aircraft along with such new innovations as radar. Because of the dangerous nature of this type of work, a lot of good airmen lost their lives in tests that failed.

This was actually my second time serving at Yokosuka. Previously, when I was a member of the Fighter 701 *Hikōtai*, I served there for two months until getting the surgery to remove my appendix. During my earlier stint at Yokosuka, I was a member of a frontline observation team with other senior pilots. We visited Taiwan, and in the Philippines, both Manila and Clark Field, to better understand what was happening in the war zone. Our group went to see the Takao Air Base (in Taiwan) after it had been bombed by a large contingent of B-29 bombers, which were operating from the Chinese mainland. The facility was heavily-damaged and there was hardly a trace of the buildings except for the exterior concrete walls. After seeing this annihilation first-hand, we were all lost in our thoughts.

While visiting the Philippines, we watched aerial dogfights that took place above Clark Field. I recall seeing enemy Grumman fighters quite frequently at this time. On one occasion, a group of water buffaloes were crossing the runway and our fighters had to wait for them to cross. Because of this delay, a commander was shot down at low altitude, right in front of us. At this stage in the war, our *Zero* fighters still performed well in combat against the enemy aircraft. However, while watching these air battles, I noticed that the Grummans had better climbing ability and speed, compared to our *Zero* fighters. After seeing this, I thought the once superior *Zero*, might not have an advantage in combat for too much longer.

Another interesting anecdote occurred after we left the First Air Fleet Headquarters in Manila. We boarded a truck to go to Clark Field. During the trip, our vehicle was strafed by an American fighter so everyone jumped out and tried to hide in the bushes along the road. While I was hiding, I noticed a large house nearby, which was very nice. As this aircraft was firing its machine guns, I was astonished to hear a piano being played (from inside this house). The music was being played as if nothing was happening; I guess the Filipinos believed it was only a matter of time before the Allies would defeat Japan.

After we completed our tour of the frontlines, there was meeting held at Yokosuka so our group could report our findings. Several senior officers and aviators attended this meeting. I presented my report from a pilot's viewpoint. I strongly suggested that we did something to better enable our fighter units to be more successful against the high-altitude heavy bombers and improve our *Zero* fighters. Modern warfare proved the importance of air power. The traditional reliance on large warships was now obsolete; this was clearly proven in the attack on Pearl Harbor. Furthermore, our Navy bombers were responsible for sinking the once mighty British battleships, *Prince of Wales* and *Repulse*. Whoever controlled the air had the tactical advantage in combat so the first step was to create stronger fighter units. Our Navy recognized this need and focused on improving our *Shiden* fighter. The goal was to create a better aircraft that was easy to fly and had better performance in order to be successful in combat against the Grummans. In the end, the Navy rolled out a "masterpiece" – the *Shiden Kai* (or improved *Shiden*).

When I returned from the Philippines, the Navy had already manufactured some *Shiden Kai* prototypes. These fighters were painted yellow to distinguish them from other aircraft. After numerous test flights, they were getting close to the final stages of testing before the fighters would be accepted into service. The best new feature of this new aircraft was the wings that had been now mounted lower in the airframe. Because of this, it was much easier to fly the *Shiden Kai*.

In early 1945, Iwashita-san while serving with the Yokosuka Kōkūtai. *(Kunio Iwashita Collection)*

In the earlier *Shiden* model, the wings were located in the mid-fuselage area. At that same time, the *Shiden Kai* fighter had a fuselage shape that was slimmer in profile. It also had a better designed landing gear and this made takeoff and landing accidents almost non-existent. Furthermore, this aircraft had state-of-the-art flaps that could deploy automatically to prevent stalling during vertical maneuvers.

Our airmen who flew the *Zero* fighter never stop praising the new *Shiden Kai*. They especially liked this new aircraft because of its safe and smooth maneuvering. It matched the *Zero* for being easy to maneuver in the air. In addition to this, the *Shiden Kai* was equipped with four 20mm cannons so it was well-armed. I felt confident that we could now do well in combat against the Grummans. We had been waiting for this fighter for a long time and I fell in love with it. However, since I was just assigned there, I didn't have a chance yet to fly the *Shiden Kai*. I only was able to check out the cockpit.

The commanding officers of the Yokosuka Kōkūtai *in 1945. Left to Right: Lieutenant Kunio Iwashita, Lieutenant Commander Masanobu Ibusuki and Lieutenant Yūzō Tsukamoto. Behind the commanders there is a* Shiden Kai *fighter. (Kunio Iwashita Collection)*

On the morning of 16 February 1945, we received an alert that American Task Force may raid the Kantō region. We were waiting in a pilots' ready room for the order to scramble. As expected, Task Force 58 came to do large scale air attacks on our mainland. We jumped into our *Shiden Kais* as soon as we received the order to engage the enemy. As our *Zero* and *Shiden Kai* fighters gained altitude quickly, we were joined by other friendly aircraft. Our fighters attacked the enemy F6Fs, F4Us and TBFs over Enoshima and Atsugi. In this engagement, we shot down over ten enemy aircraft and lost only one plane. This was the first air battle for the *Shiden Kai*. After this victory, the enemy didn't fly over the Yokosuka airspace. This was the best news we had in a long time. The pilots from the 343 *Kōkūtai* who were training at Matsuyama Base gained great confidence after this victory. However, during this mission, I was nervous because it was my first time flying the *Shiden Kai* and the weather wasn't good. Visibility wasn't that good because of the thick, dark clouds and snow. After I attacked some enemy fighters, I started to climb and gained some altitude. Just as I was climbing, a warning light illuminated and indicated there was a problem with one of the landing gears. After I landed at Atsugi Air Base, I went back to Yokosuka. One of our pilots whose fighter was damaged in combat fled to airspace over Ōfuna. Unfortunately,

A pristinely restored original Kawanishi N1K2-Ja Shiden Kai *is painted in the markings of the Yokosuka* Kōkūtai *and on display at the National Museum of the U.S. Air Force in Dayton, Ohio. (USAF)*

when he crash-landed, a local civilian defense group thought he was an enemy pilot and killed him. After this tragic incident, all of our pilots were ordered to paint their life jackets with the *hinomaru* so it was easy to identify them from the enemy.[10]

After the above mission, many of my days were uneventful. However, I did do solo missions against B-29s, which were flying high-altitude reconnaissance flights, and test-fired new armament such as bombs. Then, on 6 April, I was ordered to lead twelve *Zero* fighters that were flying to our frontlines in Kyūshū. After we departed Yokosuka and flew via Matsuyama to Kyūshū. Around sunset, I spotted the battleship *Yamato*, protected by a few destroyers, sailing southward in the Bungō Straits, towards Okinawa to do a special attack mission. I bid farewell to them by waggling my wings left and right. In the evening, we arrived at our destination, Kasanbara Air Base. We landed our fighters using the guiding lights and the base was filled with a very tense atmosphere.

By this time (in the war), the defense of Okinawa had already started and our fighter units were deployed to southern Kyūshū including: Kashima, Miyazaki, Kasanbara, Kokubu, Izumi and Kagoshima. The last phase of the *Kikusui* Operation was under way and on the day we arrived, the Fighter 312 *Hikōtai* had just had an air battle over Amami-Ōshima while going to Okinawa.[11] In this engagement, they lost over ten fighters including their commander. After I delivered the *Zero* fighters to this base, I expected to return to Yokosuka. However, the base commander said to me, "You have come here just when we need a commander because Lieutenant Kunio Kanzaki was just shot down. You should stay here for a while and command our units. Your pilots also should stay here. I will ask for permission from headquarters in Yokosuka."[12] I could not agree with him under these conditions. One of my concerns was a lack of "mental unity" between the other pilots and me. This was the most important factor because I was a new commander and didn't know my men at all. However, my Yokosuka aircraft ferrying unit ended up joining the fight against enemy fighters and B-29s; we had very unexpectedly joined *Kikusui* Operation.

One of my most memorable missions was a daytime attack, on 12 April, against the American fleet anchored near Okinawa. All of the Japanese fighters that were stationed in Kyūshū participated in this all-out attack. I was assigned to lead over sixty fighters and our formation covered a large swathe of the sky. This was the first time that I led such a large group of aircraft on a mission. The plan for the operation was this. First, our fighter group would go to Okinawa in order to draw out our enemy counterparts. Then, like playing hide and seek, we would prolong our engagement as long as our fuel allowed. Eventually, the enemy fighters would run out of fuel so they would have to land. At that moment, our bombers and other aircraft would carry out their attacks. If everything went as planned, our planes would be able to do their attacks without having to worry about enemy fighters. Therefore, our goal was to be "bait" and prolong our air battles with enemy fighters as long as possible.

After we put on our oxygen masks, we climbed to an altitude of 5,000 meters (16,404 feet). Even though we reached Okinawa, we didn't see any enemy fighters. Our group just kept circling in the air until I happened to spot dust columns on the runways below as the enemy aircraft were taking off. As planned, we kept circling over Okinawa. If I recall correctly, about ten minutes later, an experienced *buntaichō* flew past my fighter. He pointed below with his finger while waggling his wings left and right. As I gazed toward his indicated direction, I saw a large formation of enemy aircraft climbing towards us. I pulled the lever to release my belly-mounted drop-tank. After confirming that it was released safely, I gave the order to commence our attack by waggling my wings. As I put the engine into maximum power, I flew towards the enemy. This was the beginning of a large-scale air battle and I spotted an F4U fighter coming towards us. I also saw several splashes in the water below from our aircraft that were shot down. Since my *Zero* had a commander's stripe painted on it, my aircraft was attacked first; four F4Us were trying to shoot me down. Without reducing my speed, I started to descend and I kept on doing sharp turns to avoid being hit. I kept reminding myself that my goal wasn't to shoot down the enemy. I had to keep them busy so our bombers could make a successful attack. As I watched the altitude gauge, the needle quickly went lower and lower. Finally, I reached 100 meters (328 feet) and at this height, it was very easy to crash into the water. The F4Us eventually gave up chasing me. But, I can't recall how long I flew or how many maneuvers I did, to lose them. I really thought that I wouldn't survive this fight above the water. Luckily, I was successful in evading them and wasn't shot down.

Next, I tried to observe what was happening and couldn't see any of our fighters so the engagement must have been over. I felt our goal must have been achieved so I decided to return home by myself. I thought it was safe to fly back to Kyūshū via the Amami Islands. However, it was possible that some Grummans may be waiting for Japanese aircraft there so I followed a course east of these islands. On my return home, I spotted two *Zero* fighters that were flying ahead of me, but to the left, going towards Amami-Ōshima. Just as I felt they could be in danger, three Grummans who were waiting over Ōshima started attacking them from behind. Soon after this, the two *Zeros* were engulfed in flames and spiraled down to the ocean below. I continued on my eastward course and then flew northward, finally arriving back at Kasanbara Air Base. By this time, all of the other survivors from this mission had returned home so they were worried if I had survived or not. Remarkably, when I checked the fuselage of my aircraft, I couldn't see any bullet holes. I thought that the superior maneuverability of my *Zero* fighter against the F4Us had most likely saved my life.

While I was stationed at Kasanbara Air Base, I achieved something rare in combat against the (usually) superior B-29 bombers. Without any help from other pilots, I successfully shot down a B-29 with Type 3 phosphorous bombs. This bomber crashed into the Kirishima Mountain Range. In May (1945), I was finally able to return to Yokosuka. After we returned there, we did a memorial service for some of the fallen men. Even today, I don't totally agree with why an aircraft ferrying unit was ordered to fly combat missions over Kyūshū. Because of this, a lot of good men perished.

In June/July 1945, serving with the Yokosuka Kōkūtai, *from the left, Lieutenant (junior grade) Sakata, Lieutenant (junior grade) Kamiyoshi, Lieutenant Kokubu, Lieutenant Iwashita and Lieutenant (junior grade) Nakamura. (Kunio Iwashita Collection)*

The End of the War

After completing our assignment in Kyūshū, we really didn't have to fly any combat missions. The primary duties for our unit was to test and research new aircraft. Typically, I commuted to the base from my home in Kita Kamakura. In August, the atomic bombs were dropped on Hiroshima and Nagasaki then the war ended.

Reflections

I was taught some precious lessons by Americans during the Battle of Iwo Jima, my first experience in combat. As the fighting subsided for the time being, an American submarine and seaplane arrived near the shore in order to rescue pilots who were shot down in the battle. This selfless act really moved me. However, this flying boat was shot down by a *Zero* fighter. We were taught that, "Warriors should die gracefully without whining; like cherry blossoms." At that time, there was a general feeling of giving glory to death but not to people's lives. On the other hand, the United States thought a great deal about people's lives until the end.

As a fighter pilot, my military life was always

Distinguished veteran Kunio Iwashita after meeting the author in Tokyo. (Ron Werneth)

difficult and dangerous. Miraculously, I was never injured during these dangerous operations. Only a few of about eighty survivors from my class had some experiences like I had. I think, it is safe to say, I was really fortunate and must have been protected by a lucky star. This same lucky star carried me through many difficult times in the business world as a small business owner.

[1] Iwashita's father-in-law, Rear Admiral Katsuji Kondō was a very distinguished naval aviator and graduated from the Naval Academy as a member of the 46th Class in 1918. He was born on 11 June 1897 and spent much of his adult life as a Navy officer and aviator. His resume including time overseas in the United Kingdom, being a staff officer in the Air Technical Arsenal and serving as the Executive Officer on the carrier *Akagi*. All of this service occurred before the outbreak of the Pacific War. During World War II, he commanded several *kōkūtais* including: Mihoro (5 March 1941), Kanoya (1 April 1942), Ōita (1 November 1942), 381 (1 October 1943), Yokosuka (1 October 1944), and lastly, Chosen (10 June 1945). Kondō-san demobilized from the Navy on 3 November 1945 and passed away, just ten years later, on 28 January 1955. http://pwencycl.kgbudge.com/K/o/Kondo_Katsuji.htm

[2] Before leaving for the frontlines in the Pacific, Yutaka got married on 30 May 1942 and then changed his last name to his wife's family name, Ishimaru. This practice was common during this time in Japan if a family did not have a male to carry on their name. Sadly, Yutaka would never see his wife after leaving Japan.

[3] This was a "mystery" for me when I first heard this story because during this battle the destroyer *Yugiri* was not in combat. At that time, it was in Kure, Japan, getting repairs. Remarkably, several years later with the exhaustive research of Minoru Kamada, I think this mystery was solved. From the 68th Term Class of Eta Jima, there was only one man named Inoue, Lieutenant (junior grade) Tatsuaki Inoue. During the Battle of Santa Cruz, he was a Chief Gunner and could also work as the Chief Navigator on board the destroyer *Urakaze*. Yutaka's flying partner (observer) for this mission and died on their aircraft was Warrant Officer Toiichi Azuma.

[4] This *butai* was temporarily formed using a large contingent of aviators from the Yokosuka *Kōkūtai* and 27th Air Flotilla. The overall commander was Rear Admiral Teiichi Matsunaga. *Japanese Naval Aces and Fighter Units in World War II*, 234.

[5] Lieutenant Koda, Iwashita's classmate from the Naval Academy, 69th Term was killed over Iwo Jima on 24 June 1944. Lieutenant Kawakita (a 70th Term graduate) died on 3 July 1944 in the same place.

[6] 301 *Kōkūtai Kōdōchōsho*, 04 July 1944

[7] *United States Navy Combat Casualty Lists of World War II*, kindly provided by historian Jim Sawruk.

[8] Commander Ayao Shirane died on 24 November 1944 during an engagement with P-38 fighters from the USAAF 433rd Fighter Squadron. Shirane decided to crash his aircraft and did not survive. *Japanese Naval Aces and Fighter Units in World War II*, 341.

[9] On 25 October 1944 Lieutenant Yukio Seki died during his *kamikaze* attack on the United States Navy escort carrier USS *St. Lo* (CVE-63). At 1053, on this date, he crashed his bomb-laden *Zero* fighter into this ship and it sank just thirty-two minutes later. USS *St. Lo* Action Report, Battle of Samar, 25 October 1944.This is considered to be the first official suicide attack done by a Japanese aircraft. Although, throughout the war, Japanese aviators may have done suicide attacks unofficially as well. For example, if an aircraft was damaged in combat, the pilot may have opted to crash into an enemy target such as Lieutenant Fusata Iida did during the Pearl Harbor attack.

[10] The unfortunate aviator was CPO Takashi Yamazaki and this incident occurred on 17 February 1945. After he was shot down and parachuted to the ground, local civilians thought Yamazaki could be an enemy pilot and he was clubbed to death. *Japanese Naval Aces and Fighter Units in World War II*, 236.

[11] *Kikusui* means floating chrysanthemums. This was the name given for the *kamikaze* operations in the defense of Okinawa.

[12] Lieutenant Kunio Kanzaki was a 68th term graduate from the Naval Academy. He died in combat on 6 April 1945 in the Okinawa region. *Japanese Naval Aces and Fighter Units in World War II*, 401

Chapter 7

"I cannot erase the pictures I still carry from Iwo Jima."

SFC Edgar R. Fox (USMC)

I got to know World War II veteran Ed Fox when I just started getting "serious" in my interest about the Pacific War. We met online via The Battle of Midway Roundtable. The group consists of Midway veterans, historians and people who simply are very interested in learning more about this important battle. Through our correspondence together, Ed became a good buddy and supporter of my work with the veterans both in the United States and Japan.

Back in the year 2000, I visited him and his lovely wife Chris at their home in Springfield, Missouri. During our conversations together both on the phone and via the internet, it became very clear to me that he was "comfortable" talking about really anything that happened in his military career except Iwo Jima. However, after some cajoling, Ed was able to share some of his painful memories from this bloody battle with me. I feel so fortunate that he trusted me with these stories that were difficult to share. This veteran served for many years in both the United States Marine Corps and the Army. However, I feel, Ed was able to find some peace and a lot of happiness through the work he did with the children both in postwar Japan and much later in the United States. Ed would frequently send an e-mail message with some questions or information about a "school project" that he was working on. I feel blessed to call him a friend for so many years.

Early Days

I was born in 1923 and raised in Chicago. My dad was a salesman in the stockyards. My mother was a housewife, and before she was married, my mom was a doctor's assistant. We moved from Chicago to Omaha, Nebraska; right before the depression. My father was in the livestock business in the Omaha stockyards. I went to a technical high school in the same city.

Joining the United States Marine Corps

I became "interested" in the future (of our country) after the *Panay* was sunk in 1937.[1] (In response to this belligerent action by Japan), we, as a country, did not do anything. In June 1941, I joined the United States Marine Corps (USMC) and went to an USMC base at San Diego. I will never forget taking the southern route there via a coaling train. I think, it took 2 ½ days to get to San Diego. I was given $15 to get there and had a government meal ticket. Once they saw this meal ticket and noticed my youth, they "knew" where I was going. So, (on board the train) I was treated pretty well. I remember being met at the train station by a loud, boisterous sergeant. He had all of us (recruits) follow him to a cattle truck. We boarded it and went right into the Marine base.

During three months of basic training, I never saw civilian life. Basic training was all right if you did not open your mouth. If you questioned anything, look out. On a few occasions, I learned the "hard way." The first time I asked why we were going to do something, I had to go get my backpack and fill it up with wet sand. Then, I had to carry it around all day. The next time I learned about discipline was when I was smoking and flicked a cigarette butt with my finger. I did not "strip it down" and take it apart which was the "proper" way to dispose of them. This is because cigarette butts were an eyesore. So, I was sent to the PX and got a thread and needle. Then, during my off duty time, I had to go all over the post, pick up the cigarette butts and thread them together.

Daily Schedule for Basic Training

I GOT UP IN THE morning and did physical calisthenics using my weapon. I had a M1903 Springfield bolt-action rifle. During the first week, while blind-folded, I could strip the rifle down and it back together. One of the most strenuous physical drills was to hold your weapon in front of you then keep it there. This exercise would build up your muscles in your shoulders and arms. Pretty soon, physically, I felt pretty good. We had excellent food and I could never complain about it.

A few times we would do what they call a "creep and crawl." We would load up everything in our backpacks. I believe our web equipment and backpacks were called 782 Gear.[2] Our personal equipment included a helmet, bayonet, canteen, first-aid kit, a change of socks and underwear, one change of clothes, a poncho, shelter half and some stakes. Sometimes, you would spend all morning going under barbed wire with your weapon and backpack on. The pack was very light though, maybe twenty-five pounds. While on Iwo Jima, I dumped everything I had and then went forward. When you were fighting on the line, you rarely carried your backpack or much else. For example, during combat, you seldom had gear such as for sleeping, but at least, you sometimes had hot food brought to you.

After basic training, I was assigned to the 2nd Marine Division, 10th Marine Regiment at Camp Pendleton. Since I did so well with my weapon in boot camp, I was issued a BAR (Browning Automatic Rifle). It was an automatic rifle with a bi-pod. The magazine held twenty rounds so you could hold back the trigger then it went, boom, boom and boom. Next, I went on the range and made expert with it. This became my issued weapon with a .45 pistol. On Iwo Jima, I carried a 12 Gauge sawed-off shotgun and at night, it was a great weapon. When you heard something out there, you could just stick this gun out of your hole then fire it off. Then, you would not hear anymore noises.

As a unit, we went out on these forced marches across the desert for 150 miles. During one of these training operations I got bit by a rattlesnake in the neck. The only reason the doctors said that I survived was because the snake must have bit something before it struck me. There were no helicopters in those days so a jeep ambulance took me to a Navy hospital in San Diego. I stayed there for several weeks and recuperated.

War with Japan

WHILE I WAS AT Camp Pendleton, I had a knack for firearms. I also did cross-training in math and survey work. I ended up realizing that (all of this) training was preparing me to be a forward observer. I was in the midst of this training when war broke out. In fact, I was out in the field when this happened. We did not have radios like we have today. So we really did not know what was happening. I finished breakfast and was getting ready for church. Then,

I spotted a jeep that was traveling very fast. The jeep arrived and our Executive Officer was on board. He told us to pack up our gear and get back to base because war had been declared. So, we packed up everything and trotted back to base, which was maybe 4-5 miles away. It was "panic city" at the main base. There were vehicles moving around with red lights on and some sirens were blaring. I figured they must be off the coast and no information was being passed down at all other than Japan had attacked Pearl Harbor. When we got to our quarters we were told to put our weapons down and go to a briefing. Then, we were told what actually happened. Everyone was dumbfounded. Next, trucks pulled up and ammunition was unloaded. We were issued ammunition and given strict orders on how to handle it. Nobody was allowed off post and we did not load our weapons.

After this, the squad that I was in picked up all of our gear, c-rations, etc. and left on a truck. We then did beach patrols. It was total panic. For our beach patrols, we were told to load our weapons (the clips held five rounds) but not to put a shell in the chamber. We figured this was "serious" because we loaded our weapons. We did this for three or four days and it got cold out. Things also started to calm down. The Marine Corps and other military services along with our government started to focus their efforts in a more intelligent manner. Next, we were loaded on to trucks and went to North Island. There a small contingent of my company and another company boarded destroyers. This occurred before Christmas and we went to Hawaii. The scuttlebutt was that we would attack Japan. I think it took two days for the trip. We were worried about Wake Island because it was under attack. The scuttlebutt was rampant; it was terrible.

So when we got into Pearl Harbor right then we started calming down and realizing what war was like. None of us other than the senior NCOs and a few of the high-ranking officers had ever seen anything like this before. The ships were still burning and bodies were still be dragged out. This was about a week after the Pearl Harbor attack. The stench was terrible. It smelled like oil and animals that had been burnt (from all the dead bodies). Before we got into Pearl Harbor, we were met by all kinds of air cover so we were "alert" now. But whether or not the aircraft could have defended themselves I don't know. We got into Pearl Harbor but then the next morning we left for Midway. However, at that time we did not know it. On board the ship was close to a company of men, maybe 95-105 of us. I was on a flat bottom, passenger ship. It did not have a PA, but the ship seemed a bit luxurious compared to our troop transports. We were escorted by three destroyers and a PBY was flying overhead. We thought that now we were going to attack the Japanese. Then, we were told where we were going, to reinforce Midway Island and that there were more troops coming.

Private Ed fox during the time period before the Battle of Midway. (Ed Fox Collection)

The Battle of Midway

As a young private, Ed Fox went to Midway to help reinforce the atoll from a potential Japanese attack. He was a member of the 23rd Provisional Company and the 6th Defense Battalion. Midway is an atoll, consisting of two islands, Eastern and Sand. However, many people call it "Midway Island," instead of Midway Atoll. This Pacific outpost would play a major role in the outcome of combat operations with the Imperial Japanese Navy. During this battle, Japan lost four of its best aircraft carriers: Akagi, Kaga, Sōryū *and* Hiryū *along with the heavy cruiser* Mikuma. *On the United States side, the flattop USS* Yorktown *and destroyer USS* Hammann *were sunk. The land-based facilities on Midway Atoll also suffered some damage but it was never captured or put out of commission by the enemy. Japan lost 3,057 men while the United States had 307 perish at Midway.*[3] *Because of the historical prominence of this battle, numerous books and publications have been written about this engagement. Therefore, only a brief background will be included in this account as the focus is Private Fox's experiences in this battle. If you want to learn more about this battle, please see the author's bibliography later in this work for some suggested resources.*

*****

As we pulled into the dock at Midway, we were immediately shelled by a Japanese submarine. This happened while we were still at the dock. Then, our planes took off and the shore batteries started opening up. This was my "baptism by fire." The incoming enemy shells were probably three-inch ones but nobody got hit. They went over the dock and into the lagoon. Maybe, 3-5 shots were fired. Next, I learned how to construct a pill box. My position (with my team) was on the south end of the island. The Seabees were coming in and they were going to help us. Real heavy logs were put around us then they were covered with sand and sandbags. We did not live in a barracks, we stayed at our positions. Our troops laid a lot of barbed wire; there must have been none left in the US! We put spiral rolls of concertina wire on the beach (for defensive purposes). Another thing that was "taught" was how to put dynamite in boxes then pack nails around them. Then, detonators were put on the back side of the boxes. From our gun positions, we could shoot the detonators (to set off the explosives). However, sometimes a gooney bird would land and strike the side of the box. Then, it would detonate and there would be a big hole in the beach. In retrospect, I don't know how affective they would have been if the Japanese knew about them.

We had openings in the barbed wire so we could go to the beach once in awhile. It was coral sand and very white. You could walk out to the reef which was maybe a quarter mile away. Then, there was a big drop off. We thought if the Japanese landed this would be dangerous so we started laying barbed wire. Nobody was scared; we (felt) bring them on. During this time period everybody in my outfit (including me) was cross-trained on every weapon on the island. However, I don't think the pilots or ground crew did this. One of the things I did was to learn how to operate a tank. I got into this tank and got briefed how to operate it. Other people worked the guns and I thought it was pretty fun. They told me to take the tank down to the beach then turn around and come back. Then, we would fire at a target. I was supposed to go fast as if we were under fire. But, when I took off the gunner fell over and lost his headphones and helmet. I roared down there and spun around. It about threw him out of the tank. Everyone that was watching were rolling around the ground and laughing very hard. I was scared that I would run over them so I steered away. When I stopped the tank, the guy smashed his chin and nose. Then, a NCO (Non-Commissioned Officer), climbed in and told me "Do it again Fox." But, I never had an opportunity to get into a tank again.

(As a result of the allied code-breaking), about three weeks prior to the battle we were told that it was coming for sure.[4] We were notified and the brass started to arrive. Then, more aircraft and ships started to arrive. I did not think that the Japanese would be so dumb to attack the island when there would be countless destroyers ringing Midway at one time. Both cargo and troops were being unloaded as well. During the day I did not do much except go down to the range and fire my weapon. Then, during the night, I did beach patrol. Prior to this, I would sleep for eight hours. There was a barracks but we never did get up there. Instead, we were down in the machine gun positions all of the time. If there was an alert, it would take people too much time (to get out of the barracks). There was an alert every night and every morning "General Quarters" was sounded. The alert was done an hour before sunset and an hour after.

When we were told that an attack was imminent, we were supposed to not take any actions to give away our positions. Our planes were taking off from Eastern Island as there were no runways on Sand Island. I could imagine the ground crew members in front of each revetment. When the plane they serviced went by, they saluted. Then when the planes lifted off past me, I did the same thing. Whether or not they saw me, I don't know. Carrier planes landed that could not back to their ships. They either were running out of fuel or were lost. The attack was on the north end of Sand Island, where I was. I could see them dive-bombing the hangar and the buildings. One plane peeled off and circled towards our position. There was a 37mm anti-aircraft gun shooting at him. He got down on the deck so I thought this asshole was going to open up. So I got back in my hole as he went by. However, I think he was just observing. I was under orders not to fire because they did not want us to reveal our positions. After that initial attack and the enemy aircraft left, word came down to us (on what to do). If there was another attack and they came in on our end of the island, we should open fire. Then, we pulled out our guns from the pillboxes and put them in our anti-aircraft positions. Our positions were round holes that were about 5' deep with a pedestal in the middle that we would mount our guns on. There were sandbags around the holes. After this, we saw our planes coming in, smoking and sputtering. A couple of them tried to lower their landing gears and only one would come down. It was a mess.

After the attack beech patrols were increased they felt that Japanese survivors might try to make it back to the island and do suicidal things. But, that never materialized. Then, word came down that we sunk the Japanese fleet but we did not get the number of ships (which Japan lost). We also learned that an American carrier had sunk. So after this, everything just kind of slowed down. Our "limit" of five beers was lifted. We also got better food and several cases of hard liquor.

After Midway

After the battle, I probably stayed on Midway for about year and a half. Then, I went to Hawaii via aboard a ship. In Hawaii, I did some training. Then, they sent me home on a furlough and (during this time) I got orders to return Hawaii and join the 5th Marine Division. I also knew that I would be a forward observer.

(While in Hawaii) I think it was at Parker Ranch that one day a week on Saturday or Sunday (I can't remember now which day), they invited us Marines for a breakfast of steak and eggs.[5] It wasn't in mass (the whole division); you had to have a pass. I think they feed between 400-500 people at a time. That was quite amazing. Church service was held on Sundays for people who wanted to go.

I knew we were training for an upcoming invasion but had no idea it was Iwo Jima. For example, I practiced disembarking from a ship. You had to be half crazy and have a lot of dexterity to do this. You went down a cargo net and a boat would come roaring over (down below) to meet you. Just as you jumped off the cargo net, a cusp of water may drop the waiting boat down and the next thing you know, you are in the air. Sometimes, it only went down three or four feet then it met you coming down. This injured a lot of people. Other times, I would be climbing down the cargo net and there would be people in the boat holding on to the rope. Then, the next thing you would know you would be straight up in the air. When you were going over the side with your weapon and full gear, you would be probably be carrying about sixty pounds. I know that the boat crew had hooks at both the bow and stern. So if anyone fell into the water they would yell "clear" then people who were holding the cargo net would get rid of it. This way the landing craft would not smash into you (in the water). I saw this happen quite often but luckily it never happened to me.

Other times we played games similar to "Capture the Flag." This was done to see if you could get past the (opponents') positions. The people that were making the attack on your position were using tactics that they were taught. They would use diversionary tactics to try to get past you. Most of the time, hand grenades were small bags of flour that you would throw at the (enemy) troops. This training was done on the Big Island. We had a lot of intelligence work on Japanese tactics and did a lot of hand to hand tactics. Our group also learned a lot about Japanese weapons and equipment by looking at examples of captured items. So, we were able to see what we might be up against.

To keep your reflexes up, they encouraged you to take your knife down to the lava fields. This was stupid. In these lava fields were deep ravines, which were created by water erosion. In the ravines, wild hogs would travel in teams; they were viscous. We would take our KA-BAR knives down there and get somebody else that was walking above to shag them down.[6] Then, when they charged us we would try to kill them with our knives. Once you stabbed one, the others would "wise up." Some of them had pretty good tusks and teeth so you had to work pretty fast to subdue them. If there were more than three or four hogs, you had to figure out how to get out. Otherwise, you would get nabbed. It was dangerous and quite often people would get bitten or slashed then end up in sick bay. Sometimes, we would come out of there with six pigs. Then, our cooks would clean them up and we would have barbeque ribs.

Typically, we would go and eat two meals a day. The rest (of our food) was sea rations such as dried eggs, ham and peanut butter and jelly. The crackers were so hard that you could beat on them with a hammer and they would not break! I can't remember any heat tabs in those days so you had to figure out on how to heat up (the rations) or eat everything cold.

Duties on a Forward Observer Team

I DID SOME OF MY training to become a forward observer (FO) in the United States (on the mainland), but then it was completed in Hawaii where we learned how to work as a team and set up an OP (Observation Post). This position was assigned and not a volunteer one. A team was assigned to an infantry unit, like the 27th Marines. Typically, in an OP; you had a senior NCO, an officer and two or three observers. You also had protection. We had several men from the infantry unit we were to support. About the only time the number would increase would be if there was a "jump-off" planned (attack) within an hour or so. The infantry took care of us because they knew we were there for them. We located ourselves with the forward units and awaited a call from a squad or platoon leader that they had a target in front of them that needed to be eliminated. We would move up to a near location, part of the team would go left another to the right, these were called OL or OR. When we would get an OP position, it was surveyed in. You knew on the topo exactly where it was. Topos are topographic maps of the ground in front of us. In other words, they were maps of the area we are going to move, over or through, that the enemy controls presently. You also knew where the FDC (Fire Direction Center) was. Both the OL and OR would be within view from you (on each side). In training, we try to put a round into a fifty-five gallon drum that was half a mile away. In Hawaii, we would compete against other units to do this.

Typically (in combat), we would be right with the infantry. Sometimes, though, we were in front of them. When it got to the point that the OP was so far away from the advancing infantry, then we would pull out to be closer to them. We always had topos (maps) with us. Our targets could be bunkers, a tank or a troop concentration of 10-15 men. Once the target was located by the OL, OR and our center point, the compass bearing would be worked out by the FDC. The Fire Direction Center is where information from the FO is configured on a map, gun coordinates given to the firing battery for adjustment and prepare to fire when ready, often less than 15 seconds. I would say, "FDC, I have a target," and he would recognize me. Then, I would give him the grid coordinates and he would want to know what the target was. For example, if there were troops moving toward us it would be an overhead burst. Our topos were accurate to "within twelve inches." We felt so sure of our topos that we always did, "fire for effect" and rarely, "fire to adjust." This meant one round would be fired for us to move left right up or down. Normally, in the day time, it only took a very few minutes to locate and fire. At night, the same system would prevail, only this time we found the enemy by the flash of his weapon.

One night (on Iwo Jima) the Japanese did a *banzai* (suicide) attack. You could hear their horns and metal clacking. You could tell that there was a wave (of men) coming in. All these people were yelling in Japanese. There was some firing and artillery. So we explained to the FDC that had a frontal attack and that we would need everything. An emergency barrage was called in and everybody opened up on that area. We had 105s, 155s, mortars, ships, etc. They adjusted to our instructions that we received from the infantry. Star shells went off and when you looked out there, it looked like Times Square. But, in less than three minutes, it was all over.

Iwo Jima

This small island is only 7½ square miles in size, however, it was the scene of a very bloody battle in World War II. Iwo Jima was strategically important in the Pacific War to both the Allies and Japanese military forces. If the Allies conquered this island, they would have a swath of territory located only 759 ½ miles (660 nautical miles) south of Tokyo. They also would be able to construct an advance airbase that would enable fighter aircraft such as the P-51 Mustang to fly escort missions with the B-29 Superfortress bombers over dangerous Japanese-held territories. During this epic 36-day battle, 70,000 servicemen from both the United States Marine Corps and Navy faced an estimated 20,530-21,060 men, both from the Imperial Japanese Army and Navy.[7] *Spearheading the overall defense of this island was Lieutenant General Tadamichi Kuribayashi. This veteran army officer planned his strategy a lot differently than man of his Japanese military contemporaries did. He did not believe in mindless* banzai *charges where his men who die and quite often, in the end, very little would be achieved by their deaths. Instead, Kuribayashi made Iwo into an "island fortress" complete with miles of subterranean tunnels that connected well-concealed bunkers and pillboxes together. Therefore, even if American troops thought they secured an enemy position such as a pillbox, Japanese soldiers could escape in a hidden passageway and then later reappear behind the advancing attackers using this same method. One of the iconic pictures of World War II was taken by photographer Joseph Rosenthal during this battle. The picture shows the raising of the United States flag on the top of Mount Suribachi by five Marines and one Navy corpsman. Ed talks about this famous scene that occurred while he was on the island. The story of the men who raised this flag was told in the bestselling book,* Flags of our Fathers.[8]

An official United Marine Corps photograph which shows a forward observation post in action on Iwo Jima, which veteran Ed Fox describes in his duties on the island. (USMC)

Above: *An aerial view of Iwo Jima with Mount Suribachi in the foreground. (USN)*

Far right: *The USMC landing plan for the invasion of Iwo Jima. (USMC)*

Right: *Imperial Army Lieutenant General Tadamichi Kuribayashi, overall commander of Japanese forces on Iwo Jima. (USMC)*

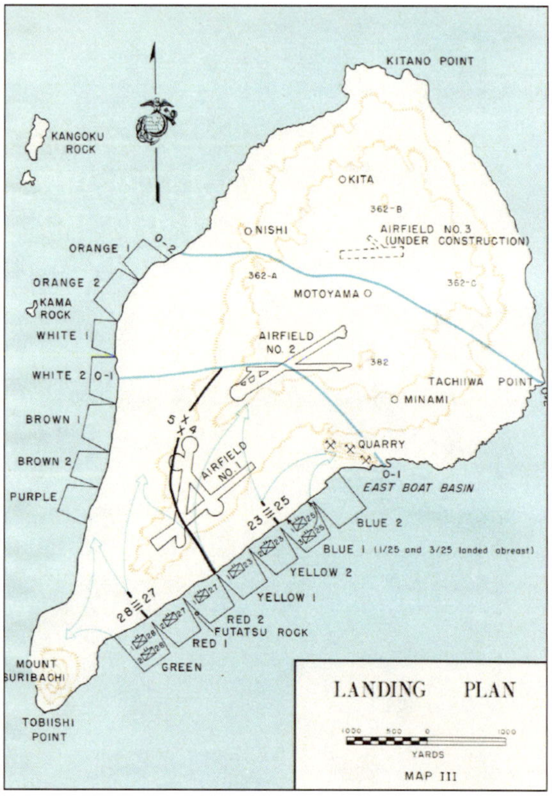

*****

I CANNOT ERASE THE pictures I still carry from Iwo Jima. Unless you were there, you can not feel the pain that is "scarred" into ones memory for the rest of his life. The date is fast approaching for the anniversary of the landing on Iwo Jima. This is the first time I have ever placed this "scar" on paper, let alone open for the public. Maybe the other members of the Midway Roundtable have helped me, by their expressions from the past is why I am unburdening now.

(We did not know Iwo Jima was our destination) until we left the skyline of Hawaii. We were aboard ship for two weeks. A mammoth convoy! I spent everyday until D-day studying the diorama and the topos that we had on board. We would hypothesize as to where they would place a gun position on the diorama or topo. I loved maps from day one and studied these topos as if my (our) lives depended on it. I had a great teacher in high school that had taught me, one could learn so many things just by learning to read and respect a map. We had aerial photos also, but they were not too clear. (In our free-time) We were playing cards and reading books until the last day.

*****

While on Iwo, Ed was a member of the 13th Marine Regiment, 5th Marine Division. He served as a member of a forward observer team and was assigned to the Headquarters and Headquarters Battalion (with no number designation). After the usual pre-invasion naval bombardment, on 19 February 1945 at 0830, the first waves of troops left their ships bound for the landings on Iwo Jima. They commenced landing at 0902 and thus would begin one of the legendary battles of the war. Every yard gained on Iwo, was earned by significant amounts of blood spilled from young Marines and Navy personnel.

19 February 1945, USMC troops on their way to the initial landings on Iwo Jima. (USMC)

Men of the 5th Marine Division hunkering down for cover on Red Beach One during the landings on Iwo Jima. Mount Suribachi looms in the background. (USMC)

*****

I WAS ON THE third wave going into Red beach. Approaching the shore line, fear is all I felt. Could I do what I was trained for? The ramp opened and we could see the bodies floating in front of us that were our own men. It was pretty rough surf and there was incoming on both sides of the landing craft. We rushed ashore only to have an explosion in front of us and a helmet hit my chest - with a head of fellow Marine in it. He wanted to be first off. When we landed on the beach there were so many wounded and dying lying in front of us. We thought we may be the first walking or crawling Marines moving inland that could get over the ridge facing us. However, not many were moving forward. Most casualties were from artillery and mortars than small arms fire. A shell hole was the safest place to be, leap frog from one shell hole to another. Yes, we experienced fear, each of us wondering if we would make our objective. What kept us going was training - habitual, repeated training from our past. We had to rely on each other to stay alive.

Iwo Jima was hot volcanic deep ash, a big knob on one end with a flat land to the north with ridges honey combed with inter linking caves. Small arms fire and machine guns did not last too long. However, there was artillery fire on us for three days. Offshore, our ships were blowing the hell out of Mount Suribachi because that is where they felt the Japanese observers where since their accuracy was so great. I don't think there was an inch of that mountain that did not get shelled.

On the beach, our tracked vehicles did have a hard time. However, for wheeled vehicles, it was suicide. You could get a tracked vehicle up to pull the wheeled vehicles in. But, this did not happen until the fourth day. (During this time of the battle), you sometimes heard men calling for a corpsman. However, most of what you heard were orders. For example, "Johnson, get your ass up here." Most of this was for getting the radios and communications gear set up. Personally, I was thinking, "How can we find out where that fire was coming from and who did it? That was our job to find them. Where we were at that time there was just infantry all around us and I could not find the other parts of my team that I was assigned to. When you got on that island and stuff was coming in, you would go where you think you could survive. It was probably two or three hours before we got all together. Our FO team was assigned to the 27th Marines, though we crawled up Mount Suribachi with the 28th Marines. The 27th had the responsibility I think on the left flank. We needed to find a location above to support them as they advanced. Once they were established on line, we came down and then joined them on their MLR (Main line of resistance) for support. We did notice that the Japanese were short on weapons. Their front line would attack with weapons but after they were knocked out, the next line would have ammunition but no weapons. Sometimes, our artillery would lay waste to the second and third columns coming forward. So it was up to the infantry to get the first column. Some of the Japanese in their second wave attacks had whistles, trumpets and drums. They were trying to harass us.

On my third day on the island, we were supporting the 27th Marines adjusting for 105s (howitzers). At dusk we began to take heavy fire on the line. Our job, as well as our OL and OR were to attempt to locate the Japanese artillery by their "tube flash." We would lay in our glasses on the flash and then call the bearing to the FO; from there we could create a triangle back at the FDC. In most situations like this the command was "Fire for Effect" or we would ask for an emergency barrage. If one requested an emergency barrage "ALL" guns, left, right, or at sea, if possible, would fire on the target you asked for. One cannot imagine how the adrenaline is pumping through your system. Flash bangs all around you. The ground is shaking from the concussions of the incoming. Some are air bursts and shrapnel chunks bouncing like rubber balls all around you. The din is unexplainable; the "on the way" shells going over and the Japanese shells landing nearby. One tries to become as small as a grain of sand in all of this "violent disorder" but still do your job. Selected members of the 27th Marines were doing theirs; protecting us so we could spot the Japanese guns. They had a protective ring around us and in so doing, suffered most from the incoming fire. We were able to silence the guns after having a severe "counter battery exchange."

Once the incoming ceased, we began to assist the wounded. A Marine pfc. near me called for me to help him. He said he could not walk. Darkness had fallen and we were now concerned about frontal attack. We laid quiet, listening for the rush, and all I could hear was this Marine asking for me to help him. I crawled from my position and joined him to my left, only ten or twelve feet away. I remember his expression as I pressed my face to his to whisper to be quiet. His arms went around me like a vice grip. I had to make a quick decision, call out for a corpsman or carry him back a few yards to where I knew the corpsmen were working. He just grabbed me and said, "Mother." As hard as I tried, I could not get him to release me. Off we went to the rear, scared, that I was going to get it in the back while carrying him. When I attempted to lay him down next to other wounded a corpsman had to help me get loose from his grip; it was then I noticed he had lost both legs at his thighs. He had died as I carried him to the aid point.

There was more incoming fire as I crawled back to my FO position. Hunkering down again as I attempted to use my binoculars, this time the fire was more intense, the glasses were knocked from the tripod. It would be suicide to try and remount the glasses. I just rolled into a ball, covered my head with my hands and wished all this shit would go away. Two or three minutes and it was all over - a deathly quiet. Waiting again for the rush. It came but not near my location. The attack was fought off by the 27th Marines. We then tended to our wounded again. However, this time when I returned to the aid point, I found that they had taken a direct hit. All of the wounded along with the two corpsmen were killed. I cried. I became angry. I wanted to take the most horrendous revenge one could ever think of against the Japanese on that island, now, not wait till morning. I went back to my FO position and waited for another mission. I have never forgotten that night. It all took place in just a few minutes. What a waste. The disadvantage of being hateful or holding hateful emotions is that they can lead us in the wrong direction later in life.

(Around this time), I got some hot shrapnel in my rear end. I did not know what was wrong and it was uncomfortable. The corpsman lowered my pants and put something on it. He wanted to take my name for a Purple Heart. But, I said no, I don't need that, a Purple Heart for a blister on your ass?

When the flag was raised on Mount Suribachi I was only a few feet away, using my spotting scope to the north of the island. We said, "Look at those gung ho bastards. Look what they did. Who was smart enough to carry a flag in their backpack? I guess I could have assisted with the flag. Instead, I was looking for gun flashes; just twenty feet from historical event. We set up an observation post on top of the mountain. It always was hot, smoky and stinky. The whole island was hot. There was a cave on the mountain and inside was a hospital. I got in there and saw about 20-30 bodies. I also found a sword but I sold it after the war.

In general, the Japanese were not really entrenched; they were pockets of them. Most of it was automatic fire. Only at night would you find that the infantry would get up and advance towards you. They were maybe company strength. Very few times they were organized or in large groups. It was about the fifth or sixth day you could tell that the advances by the Japanese were becoming desperate.

They had tunnels that would go for long distances and there would be a room off to the side. This room could be a holding area for men or a storage area for equipment. I guess the entrances were camouflaged well at first. But once the artillery got going, they would destroy the entrances. There would just be a hole so we knew that there was a cave there. However, in some instances, as we would advance the entrances were on the opposite side (of the area), so we did not see them. We would get information from our observation planes. Then, we would bring in the mortars. A 155mm howitzer would just raise hell. They loved those types of targets and they would fire on them from the opposite side of the hill.

As the battle progressed, there were fewer targets. You could spend two or three days and have no targets. We would fire star shells with parachutes all night long. So the frontlines would be illuminated all of the time. Once in a while, we would think that something was out there so the infantry would open up with machine guns. If there was return fire, then we knew there was a target. I think only twice I witnessed a "counter-battery" (artillery vs. artillery) engagement twice. I would stay right there in my hole. The enemy did not run over me. They never got that close. We told (the men manning our artillery pieces) to drop their range and bring it in closer. Sometimes, they would get too close. Once in a while our people back there would panic and not put enough bags of powder behind the projectile (with a proximity fuse) to get the proper distance. Because of this reason or if we had a "short-round," they would land near us. However, when you had counter-battery fire, you could not tell who was who.

On the western side of the island in a lowland area there were land crabs that would scare the hell out of you. At night, they would come out and crawl all over the place, into your hole, over you while in a prone position when you didn't dare move. The crabs were really demoralizing; they were in the hundreds, worse than ants. You could not tell if the crabs were coming or the Japanese. If you had any type of cover over your hole, they would just run over it. You would be just about dozing off and five or six of them would run over the top of your poncho and wake you up. It also was not uncommon to have the sulfur fumes come up out of the hole around you. Your clothes stunk and everyone smelled like sulfur. If we were not getting fired on, we would not get into our holes. Instead, we just lay on the ground. The holes were terrible. When I ate (maybe on the fourth or fifth day), a company mess sergeant had heated up c-rations for us. When you spent some time in the infantry, they would take some portions of the men off the line to eat.

On one occasion near the end of the battle, the Japs were coming in and out of a cave that we were watching. So we blew the hell out of it. Nobody moved for two days. Then, we moved forward and this cave had rocks not blocking the entrance. It was sealed but we knew somebody was in there and alive. We took the entrance apart and I went in with a weapon and had grenade. There was help behind me. We found this command post area but we were scared to touch any of the obvious things. There was a tunnel there. So we started going down it but you could hear noises down below. Then, we stopped, turned the flashlights off and waited. Somebody was coming up the tunnel so I pulled the pin out of my grenade and went against the wall. It went off and you could hear yelling. We the put some more grenades down the tunnel and it got quiet. Then, we started going down the tunnel. Actually, what happened was that from another hole some Marines were coming in. They heard some explosions so they threw phosphorus grenades in. But, before we got out of there we were breathing these fumes. It was a whole tunnel network. We got smart and got out of there. We waited until the next morning and put some lights in there.

As far as my thoughts on Iwo Jima, I have never seen or read any of the movies or books on Iwo Jima. When I left the island I acquired a very bad drinking problem, I just could not get those memories out of my mind. Liquor helped too well. I have never revealed my past emotions as I have today, but the incident above is like a few others I witnessed. Just think of the poor guys that died "alone" and the ones that suffered before help could reach them. Even though this took only less than twenty minutes to log in my memory, some fifty-five years ago, I can still see that shell hole in the aid point (which was hit).

*****

Ed's personal account gives a good glimpse of what is was like to fight in this 36-day battle until the capture and occupation was formally declared "completed" at 0800 on 26 March 1945. The number of men (from the United States Army, Marines and Navy) who died in the Battle of Iwo Jima is considered to be 6,821. This number includes both KIAs and soldiers who later perished as a result of battle wounds. The total includes three officers and forty-three enlisted men which to this day are still considered missing so they are presumed dead.[9] An additional 19,217 men were wounded in battle along with 2,648 other soldiers who suffered combat fatigue. However, the sacrifice of these men was not in vain. As a result of the capture of this island, 2,251 B-29 Superfortress bombers flown by 24,761 aviators had a place to land if their aircraft was either damaged or low on fuel. Undoubtedly, if Iwo Jima was still an enemy-held territory, many of these airmen would have not survived these dangerous bombing missions over Japan.[10] Based on the historical military reports, Japanese losses for this small piece of territory were 19,977 killed and 1,083 prisoners.[11] While reflecting on these sober numbers of men lost it is clear to see that the Japanese combatants preferred to either die in battle or commit suicide instead of becoming a prisoner of war. As for General Kuribayashi, he was deemed to share the fate of so many of his subordinates and would never see Japan or his family again. However, the exact cause of his death such as ritual seppuku *(suicide) or combat is not clear to this day.*

After Iwo Jima

Eight or nine USMC guys including me went aboard an LST and our job was to take it back to Norfolk, VA. The top speed was about 8 knots. We figured that the war would be over by the time we got half way across the Pacific Ocean at that speed. I could not cook and did not want to learn. I also told them that I would get sick and throw up in the engine room. So, they put me on the bridge as a helmsman. All I had to do was help them keep the ship on course. A few times it got so dull up there; I would be the only one on the bridge. An ensign was the commanding officer and he was younger than me. When we got near Honolulu, we got a Navy crew on board. We stayed there and what was left of the 5th Marine Division came in. Then, we trained for the anticipated invasion of Japan. We figured that it would be a "rough road" during the invasion of Japan. But, the atomic bombs were dropped and it saved millions of lives on both sides.

*****

Once combat operations with the nation of Japan officially ended, Ed had one assignment – to be a part of the occupation troops who helped bring stability to the war ravaged country.

The Occupation of Japan

A pretty good amount of troops went to Japan. But, we were still cautious about what might happen when we took over the occupation. We landed in Sasebo and got off the ship then went further inland. I remember the Japanese people on the side of the road would bow as we passed with the American flag. After the flag would pass they stand in an upright position. Nothing but "respect" for our flag impressed me. We boarded trucks and went to various places inland. I was eventually assigned with several other people to Ōmura. I also was a bodyguard to medical teams researching the effects of the atomic bomb on victims at Nagasaki. It was the worst picture you could imagine. There were still burnt bodies under the debris and the stench was intolerable. One had to breathe through your mouth. Then, you had flies to contend with. The population would turn away when I or the research party would approach. When I would speak with the citizens through an interpreter the Japanese would bow, look down and never look up while we questioned them. But, they were always polite and some offered me tea and rice. The people were not starving. However, there was some illness but this was due to dirty water and spoiled food. I never gave my food away, though I often invited a family to share my food - but never the same family twice. (While at Nagasaki), I took note of the equipment and industry there. An awful lot of stuff came from the United States many years ago such as some machines and lathes.

The destruction at Nagasaki was incredible. This picture was taken over one year after the dropping off the atomic bomb during an allied study of the affects of this new weapon of war. (NA)

Some of the local Japanese were reluctant to say what they did. Maybe they had done something for the war effort and thought that they would be punished. On one occasion, there was a Japanese military veteran in this town that had died from his war wounds. I attended his funeral. This impressed the Japanese that an American would show his respects and I heard this later on.

The children in the area went to a pretty decrepit school. I noticed that they just had some jump ropes. I went to our recreation area but they did not have enough baseball bats and balls that I could take away. So, I appealed to the Navy and got some baseball equipment off a ship. Then, we began playing baseball at the school. Every kid from the school turned out on the first day. I think I always wanted to be a teacher and I went to school (in Missouri) to be one but I came down with a lung disease. Because of this ailment, I never received my degree. The Japanese adults could not be swayed in their thoughts. Instead, much of my free time in Ōmura was spent with the kids at school. The kids' minds were open to learn. I played baseball and obtained some baseball equipment for them. With an interpreter I often sat with the teachers and kids talking about the United States and my home. I had pictures from home that I gave them. They in turn told me about themselves, their parents and sometimes about the war. I never asked about the later topic, they volunteered their feelings. I was assigned in Ōmura for about a year.

Ed during the summer of 1946 while on leave in Omaha, Nebraska. (Ed Fox Collecion)

However, by this time, I had enough horror and bad memories. I wanted to go to the land of ice cream, good beer and women who I could feel comfortable with.

Return to the US

After my time in Japan, I went back to Camp Lejeune, North Carolina. There I became a manager at a movie theater and had several people working for me. But, who wants a person in civilian life to adjust artillery? So, I got out for a few months (but then enlisted in the Army). I enlisted in the Army because I never wanted to be a forward observer again, too many sad memories. Secondly, in late 1948, the Army offered me schooling with the Counter Intelligence Corps (CIC) at Fort Holabird, Maryland. Then, I was sent to UCLA (University of California, Los Angeles) for a six-month crash course on cinematography. I was assigned for a few months on civilian motion picture sets as an observer, then, I was sent to the US Army Pictorial Center in Long Island, New York. Within a few weeks of my arrival, I was interviewed by a State Department representative for a classified mission involving the recording by film - the surveying of the possibility of the DEW line across the Greenland Ice Cap. I received Top Secret clearance and my military career took off with this assignment. The project had a code name "Pole-Ax." Included in the contingent was Admiral Byrd's son, Sir George Hubert Wilkins, an England representative, who was a distinguished Australian polar explorer.

The Russians would harass us by flying overhead and our guys would chase them away. You had to be careful when filming there because it got down to 88 degrees below zero. The assignment on the cap proved to be a near impossible feat. I took my camera apart and there would be no lubricant whatsoever. Camera equipment had to devoid of oil with a substitute of graphic powder. Batteries would not hold up so everything was hand-cranked. The black & white film, in hundred foot rolls, became brittle in the extreme temperatures. Film canisters had to be carried in my arm pits and inner pockets close to my body. A nurse down at Thule Air Force Base donated a bra that allowed me to carry a few more cans of film. I was the blunt of many remarks and razzing. With the start of the Korean War in late June 1950, this ended my Greenland Ice Cap mission. This was okay with me because Greenland was a cold place.

Korean War

I was a member of the 167th Signal Company and got my orders from the United States Forces Headquarters. So, I never had any problems. When I felt I had some important film to get out, I went to the nearest airstrip. There I would request a courier bag with a diplomatic pouch that had instructions where

SFC Signal Corps Combat Cameraman Ed Fox covering Republic of Korea Armed Forces in action during the Korean War. (Ed Fox Collection)

to go. Then, a pilot could take it to Japan and within a few hours they would be processed. I filmed supply lines; for example, the Red Bull Express was a major supply line in Korea. We had aid stations up and down the highway. I also covered the peace conference in Panmunjom, Korea.

Postwar Days and Family

AFTER MY SERVICE IN the Korean War, I went back to the United States mainland. Then, I participated in the nuclear testing done at Enewetak Atoll (part of the Marshall Islands in the Pacific) where they put a ship at ground zero. We put our cameras on it and listened to the count down. At five seconds before zero, we pulled down our goggles. When it went off, there was a tremendous shock wave that went up straight in the air. I operated the high-speed camera. After this, the testing continued in New Mexico (United States mainland) where I did nuclear training films. For example, how to take a nuclear device apart and then, how to put it back together. These films were made for all branches of the service (Army, Navy, Air Force and Marines).

Next, I went to Germany and did the same duties where I took all kinds of motion pictures and photographs. My wife and I met in Europe. Chris was teaching in a military high school for American dependants in Kaiserslautern, Germany. We got married in 1960. Our son was born in Germany. We took a one month honeymoon in Europe, visiting many countries. We have two children, a daughter who is forty-four years old (unmarried) and a son who is forty-six. We also have two grandchildren by our son's marriage.

Ed at Sukchon, north of Pyongyang, Korea, just after his jump with the 187th RCT. He did a practice jump prior to his combat one. Some of his film footage was used for the Army documentary series The Big Picture *while other portions were used for other military purposes. (Ed Fox Collection)*

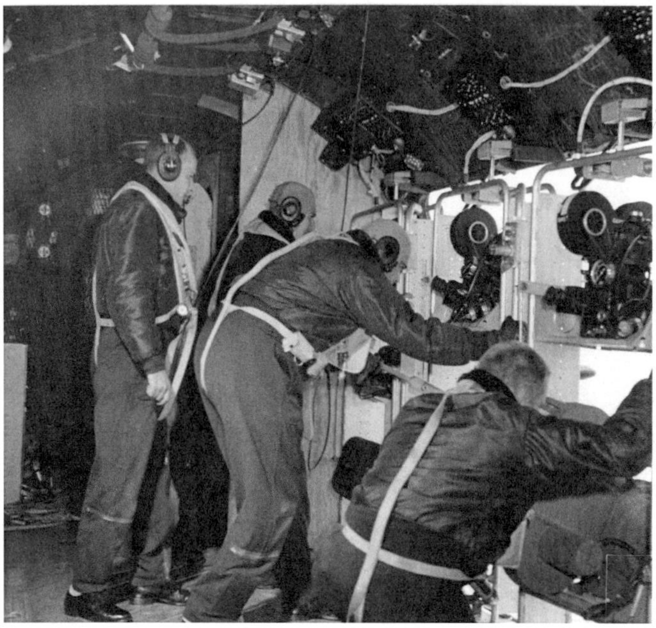

The filming of one of the Enewetak Atoll nuclear tests with several different types of cameras to document this lethal new weapon of war. (Ed Fox Collection)

Ed Fox in 1964 after over two decades of military service with both the United States Marines and Army. His service covered everything from the battles of Midway and Iwo Jima through the Korean War then the ensuing Cold War era. (Ed Fox Collection)

Following Europe, I was sent to Springfield, Missouri, as a counselor recruiter. I did this until 1966. Next, I worked in law enforcement until retiring. After retiring, I did volunteer work for the US Forest Service and managed a work station for them. My wife obtained a teaching job with the Springfield school system were she taught for twenty-seven years. I often became involved in her projects and field trips. When she retired, I continued to assist the teachers at Holland Elementary School for close to fifteen years. The school was a short walk from our residence.

Reflections

How did I feel about the Japanese, then and now? My job was to destroy the enemy before he could do harm to me. It was them or me. I was too young to realize the consequences. I was invulnerable, cocky and held my own against any odds, scared yes; fear never. Mentally today? What a "stupid shit" I was. But, that "don't give a shit attitude" is what carried me though several close encounters. Today, some of my friends are of Japanese decent. I have had students that were Japanese and we became very close. I have invited Japanese friends to our table.

Regrets for participating in a war? Not World War II. We were attacked and out to teach them a lesson. Everyone that I went to high school with never came back. They were in all branches of military service. Korea was a just cause. In Vietnam, we should have done things differently, kick out the media, shoot at the politicians, scare the shit out them that came over to see how Ma Ma's little boy was being treated. Let the military do what they are trained for. Our higher ups, the president and congress did not know what to do. They had bad advisors. Don't let our politics and military get to such an extent of disrepair and head in the same thinking like we had in the beginning of World War II. We had such poor equipment. The politicians did not know what the hell was going on. Our intelligence department was not up to date. Politicians were against the funding of things like that. I think the military should be able to do what they have to do. They should not have to justify how they spend their money, especially to get information. The congressmen and senators have a better healthcare plan than veterans do. They also have a better retirement plan. In retrospect, as for me, would I do this all over again? Yes, with very few changes.

[1] The USS *Panay* was sunk on 12 December, 1937 by belligerent Japanese military forces in China. The loss of life was two United States Navy men and Italian journalist Sandro Sandri. Eleven other Navy officers and men were seriously wounded and so was *Collier's* magazine correspondent James Marshall. The gunboat was being used to evacuate American officials and civilians from Nanking, which was under attack by Japan. However, the United States had informed the Japanese government of their plan in advance but the *Panay* was still attacked and later sank. This incident almost started a war between the two nations but the Japanese government issued a quick apology. Japan followed this apology by an offer to pay reparations for both the loss of the USS *Panay* and casualties.

[2] 782 Gear is the acronym given for the individual equipment issued to a Marine. The acronym comes from the number of the form (Marine Corps Form 782) which the soldier signed when the gear was issued to them.

[3] The source for the Japanese losses was: Jonathan Parshall and Anthony Tully, *Shattered Sword*, p.524. The American losses were derived from: "The Battle of Midway", Office of Naval Intelligence.

[4] The importance of cryptanalysts and their successful breaking of the Japanese codes undoubtedly played a big role in this battle. Throughout the spring of 1942, there were Japanese messages that stated the next objective for their military would be "AF." However, it was not clear what "AF" was until a fake message from Midway Atoll was sent out which said there was a lack of fresh water due to a damaged distillation plant. This phony message was intercepted by the Japanese and the enemy confirmed that "AF" was indeed Midway by stating it needed a supply of fresh water.

[5] From late 1943-1945, part of Parker Ranch was called Camp Tarawa and used primarily by the United States Marine Corps for training. The ranch is located on the Big Island of Hawaii. Over 50,000 men including Marines, Navy corpsmen and Seabees did training for both the battles of Iwo Jima and Okinawa there. It is still there and remains a thriving ranch.
[6] KA-BARs were the popular and quite affective fighting/utility knives carried by members of the United States Marine Corps in the Pacific.

[7] *VAC IntelRpt,* Encl A.

[8] It was written by James Bradley, the son of the late John Bradley, who was the son of John Bradley, the Navy corpsmen who participating in this historical event. In 2006, this book was adapted into a film of the same title and directed by Clint Eastwood.

[9] Lt. Col. Whitman S. Bartley, USMC, *Marines in World War II, Historical Monograph, Iwo Jima: Amphibious Epic*, Appendix III, Casualties.

[10] Office of the Assistant Chief of Air Staff, Intelligence, Washington, D.C. "Iwo, B-29 Haven and Fighter Springboard," *Impact,* Vol. III, No 9, Sep-Oct45, 64.

[11] Patrick K. O'Donnell, *Into the Rising Sun*, p.227 for the estimated number of Japanese killed. Prisoner of war figures were taken from the original 1945-dated United States Army reports.

Chapter 8

"When we came back, there were bodies all over the place."

Major Merton P. Hansen (USMC)

As I entered the doors of St. Paul Presbyterian Church (Johnston, Iowa) with World War II United States Marine Corps veteran Mert Hansen accompanied by his wife and family, I received a warm welcome from everyone there. We came to attend their annual fall Thanksgiving dinner and I was a guest of the Hansen family. Mert is one of the Elders at this church and helped oversee its construction during his postwar years. As he grabbed me by the hand and gave a personal tour of his house of worship, I could see how proud this gentleman was about having been involved since its early days. Everyone there made me feel so welcome even though it was my first visit to their parish.

However, my main reason to visit Iowa was to interview one of the last men alive to fly dangerous, low-level reconnaissance missions over Iwo Jima. The purpose of these flights was to spot enemy troops and their defensive positions on the battlefield below. Flying with him was a forward observer (like veteran Ed Fox who was profiled earlier) and they would plot this vital information on a map then send it to their counterparts on the ground, in the air and at sea. Therefore, this would enable them to do more accurate attacks against the Japanese using all the tools of war they had such as artillery fire or air strikes by support aircraft. Remarkably, these brave Marines typically flew between fifty to one hundred feet above the battlefield, in an unarmed, fabric-covered Stinson OY-1 (L-5) Sentinel. Even the Japanese troops knew this and sometimes waved at them from the ground! I met Mert and his family just by chance and good fortune while attending the well-known yearly EAA AirVenture air show (Oshkosh, Wisconsin) during August 2013. After meeting this humble gentleman, I knew that it was important to properly interview and share his incredible story. Special thanks to Ikuo Shimizu, Stephen Hansen, J. Michael Wenger and Duncan Cameron for their assistance in helping write his story. All of the dates shown below in Mert's story were derived from the official wartime documents in his United States Marine Corps Service Jacket. Sadly, shortly after I interviewed this amazing veteran, both he and his wife Jeanne passed away but they both were able to see this story and verify that everything was accurate as per Mert's amazing memory.

Early Years

I was born on 2 June 1921 and had the opportunity of growing up during our worst economic depression in our nation's history. We lived on a farm in southern Minnesota. During the 1930s, while I was in my teens, there was a tremendous drought and most of the farmers in our area lost their farms. In our local country school, there were probably twenty families who lived on twenty different farms. These farms were of different size land, which ranged from maybe twenty acres to two-hundred acres. During the Depression, only four or five families were able to keep their farms. Because of this, we never knew would have to leave our school. In 1939, when I was a senior in high school, my father lost our farm. When I graduated, I wanted to go to the University of Minnesota to study agriculture but there was no way for me to get there. So, I worked for a dairy herd improvement association, testing butterfat in milk and advising farmers about improving their herds.

My job was a lot of fun but when war broke out, I went to Minneapolis to the enlistment station because I wanted to fly. During those days, I would be working in a field and when an airplane would be flying you could hear it for miles away. Then, I would stop what I was doing and watch that plane until it was out of my sight. When I did this, my dad used to get sort of disgusted with me. I thought, gee, what a deal it would be up there and see all that country. These thoughts would go on and on. Then, I would even have dreams that the whole western sky would be covered in planes. I later got to see this at the Nimitz Day Parade in Washington, D.C., because I didn't have to fly that day because they didn't want our "little" planes getting in the way.[1] Just beautiful though.

The Army said to become a pilot, I had to have at least two years of college, but I only had high school. However, they said that I could enlist in the Army and if I qualified as time went on, I might have a chance to fly. So, I went to the Navy and they had the same story. But, just as I was about to leave, a Navy Chief called me over to him. He said that the Navy had a "test" which they wanted to try out on some of the men without college experience. The Chief went on to say that if I passed this test there was a possibility that they would lower the requirements for taking flight training. So, I ended up taking the test and stayed there a couple of days to get it done. It was a challenge, for example, the mathematics, which I never studied before. I sat there for quite awhile. The Chief said that I passed this test and would be allowed to fly if the requirements changed. After this, I went home to our farm.

Mert joins the Navy

On 2 June 1942, while I was milking a cow, they announced on the radio that the Navy dropped its college requirements to become an aviator. So, if you passed their test, you would be eligible for flight training. At that time, if you were under twenty-one years old, you needed to have a parent's signature in order to be allowed to take flight training. I had put these papers on my father's desk but he never signed them. That morning my dad asked me, "Are you going to enlist?" My parents tried to discourage me from flying because at that time, seventy-five years ago, it didn't look like a "safe" occupation. My dad said that he had to sign those papers but I said, "Not this morning." He replied, "Oh, that's right, you are twenty-one years old now." I went to that enlistment station and the Chief was wondering what I was doing there. I said that I came to enlist but he replied that the Navy hadn't changed its requirements yet. I told him that a radio announcement this morning said that this changed. The Chief was surprised and said that he had to check the morning mail. After this, he came back with a big grin on his face and said that I was in!

*****

MERT ENLISTED IN the United States Naval Reserve on 3 June, Class V-5, with the rank of seaman second class. In this program, he was supposed to serve four years with the Navy, but in his case, the young veteran would later transfer to the United States Marine Corps.

*****

Over half of the people that I trained with (in the Navy) were college-educated. However, growing up in a small town of only 1,500 people, the only "college people" were veterinarians or doctors so I really didn't know any of them. Therefore, I was scared to death when I got in with those guys. But, I think, this factor also helped and prompted me to study very hard. My biggest problem was doing Morse code because I had no rhythm. It seemed to me that the guys who were good musicians learned this the fastest.

The Navy said that the first thing I could do was take the primary CPT (civilian pilot training) at the University of South Dakota. I wouldn't be paid but they took care of my room and board along with all of my expenses for flying. After completing this, they suggested that I take the secondary course, which lasted almost two months. It was really fascinating and enjoyable for me. I got to solo, do aerobatics and talked to like-minded others about flying.

Next, from 25 November 1942 until 26 March 1943, I went to the University of Iowa (Iowa City, IA.) for my physical training. I recall the difficult swimming tests that we had to do. For example, I had to swim the length of an Olympic-sized pool using one stroke method, and then go back using a different stroke. Then, I still had to do it one more time using a stroke of my choice. I was scared to death of water and even had to jump off a thirty-foot high tower. During one of these jumps, I had to be fully clothed but I could take off a few pieces of clothing if I wanted. Another exercise was to jump into the pool and swim fifty feet underwater. I worked and worked on all of this. Then, on a Sunday afternoon, I went over there to take my test. Everything went so well with my first test that kept going until I passed them all that day. My three roommates were cheering me on so I couldn't let them down. They helped me a lot during my biggest challenge in the physical portion of my training.

After this, I went to Minneapolis on 26 March until 6 May 1943. There, I did flight training using a Stearman biplane. Prior to this, I flew an Aeronca in primary (CPT) and a Waco during secondary (CPT). The Stearman was a lot like the Waco biplane and had an open cockpit. It was February and very cold for flying in an open-cockpit aircraft. They issued us heavy leather, fur-lined flight suits and long gauntlets so you couldn't wear a wrist watch. This was the time of "dollar watches" and we wore them on shoestrings around our necks. Not many clocks in our aircraft worked and they were pretty strict about getting back on time because if you were late, the next guy couldn't start on time. It was sad to see the guys who would get sick in the air. We wore these chamois masks because our faces would freeze if we didn't. Anyhow, these pilots who were sick would take their masks (when they landed) and peel frozen vomit off their faces. There were a lot of washouts because it was pretty hard to learn how to fly during that age. Fortunately, for me, I already had about a hundred hours of flying time before this course so I had no problems there.

One of the Stearman biplanes used during Mert's flight training. Remarkably, even today, these beautiful aircraft are still popular and frequently can be seen at airshows. (Mert Hansen Collection)

Then, from 8 May until 11 September 1943, I went to Pensacola, Florida, and started flying the SNJ for a series of practice flights. From that aircraft, I transitioned to the SBD Dauntless dive-bomber. There was one field that was outside of the main field and it was set up for dive-bombing training. I thought that the SBD was a lovely plane. During night flying, you weren't allowed to use any lights so the only way to check your position was by looking at the exhaust fumes of the guy flying next to you. The Dauntless was a good plane and probably the last of the "good" dive-bombers. I recall knowing a few days before graduation that I would be a Marine pilot (not Navy). I told them that I would like to join the Marines if they needed me. There were six of us who joined the Marine Corps in my graduating class. Because of my swimming ability, I thought that the Marines would be a better deal. However, most of my USMC buddies ended up serving on aircraft carriers, just like their Navy counterparts. For example, on the carrier that I was serving on at Iwo, the dive-bomber squadron pilots were all Marines.

*****

On 1 September 1943, Mert was appointed a second lieutenant in the Marine Corps Reserve. Nine days later, on the 10th, he accepted the appointment and took his oath of office at the Naval Air Training Center, Pensacola.

Marine Corps Days

The day we graduated, they announced after graduation, that all the Navy pilots could go on leave for a few days, but we Marines had to report to this auditorium. The six of us walked in there, but all together there must have been about fifty men present. The others were corpsmen, mechanics and other enlisted Navy personnel. The latter men had gone through the Navy training and transferred into the Marines. The first guy who talked to us said how "superior" we were to be selected for the Marine Corps and we could do so much more (than other men). I thought to myself that this was a bunch of crap as the other people in my class were flying the same aircraft that I was. How could I be superior to them? However, this was good because you were getting ready to kill someone and you don't want to love them; it was a pretty hard thing to do. This was our first lecture and we always had lectures every time we changed bases.

Regarding the Japanese, they told us about how cruel they were to their prisoners. We were led to believe that they were cannibalistic. There wasn't anything worse than a Jap. During the war, you never said "Japanese," they were just Japs. It took me a long time to get over these preachings; the Marines are great for that. I think back now though, and still believe it was done for a lot of good reasons.

The training (for both the Navy and Marines aviators) was the same except when I graduated they sent me on 16 September 1943 to Daytona Beach, Florida, where Marine pilots

Mert as a young Marine. He initially trained to be in the Navy but later transferred into the Marine Corps. (Mert Hansen Collection)

were doing dive-bombing with SBDs. We had three months of intensive training practicing dive-bombing technique, skip-bombing and gunnery. This training was completed on 16 November.

A group photo of Mert and his fellow aviators who were training in the SBD Dauntless dive-bomber at Daytona Beach, Florida, in the fall of 1943. Mert is standing in the back row, second from the left. On the back of the original photograph, there were nicknames used for many of these aviators such as "Fearless" or "King." (Mert Hansen Collection)

After this, on 2 December, they sent me to Cherry Point, North Carolina. When I walked into this room where the squadron was, the colonel introduced me to the group of pilots as the only guy who flew an SBD and that I would show them what to do. I did about three or four days of that. One time, I took off and couldn't get the plane out of low pitch. I went about twenty miles before I got enough altitude to come back. The mechanics said that this aircraft came up from Florida and they forgot to get the oil changed in the hub. It had hot weather oil instead of cold weather oil so they fixed that. Then, every time I showed the fellows how to react during an overhead gunnery run, the motor would cut out when I flew upside down so I would have to restart the engine. I kept bitching about this and found out that they couldn't get any 100 octane gas which the SBD needed. So, I quit doing position changes and stayed flat level so she ran all right after that.

One time, I got so disgusted that I took this SBD with a passenger and put it through one of the worst flights I ever did. I probably had to restart it about a dozen times and saw that this was bothering my passenger so much; I enjoyed the flight that much more! When I landed, he said that he didn't want to ride with me again. He actually "messed" in his pants and when he shook his pant leg, some came out.

I talked to this colonel about being transferred and he tried to talk me out of it. He said, "You are the only dive-bomber pilot I got and I have to form a squadron out of those thirty-four pilots. I said that is why I wanted a transfer and didn't want to go through that all again. He tried to change my mind but in the end, gave me the transfer and I said thank you. Just as I was leaving the colonel said that he checked about this transfer himself but they wouldn't take anyone above the rank of a captain.

In late December 1943, I went to Quantico, Virginia, and was assigned to my new squadron, Marine Observation Squadron 5 (VMO-5) on the 29th of that month. The interesting thing was about the fourth or fifth day as a bunch of us were standing around when a familiar person walked up to me. It was Jerry Lockwood; the guy that I have given the "crazy" SBD ride to! With his hand over his mouth he said, "If you say anything about that last ride I had with you, I will kill you." Of course, he didn't want that story out and I promised to keep quiet about it. I kept quiet about this story until we came back from overseas. Then, I asked him for permission before sharing the story of our ride in a SBD. By that time, it was okay because we were all good friends. After the war, he became a harbor master in Charleston, South Carolina. Unfortunately, he passed away a few years ago.

Next, in late April 1944, we went to Camp Pendleton, California, where we trained together for several months. Then, on 30 August, we boarded the USS *Hancock* at San Diego (California) and it took us to Hawaii. There, we (our squadron) trained on the Island of Hawaii (the Big Island) where the 5th Division was. Our training was done at Camp Tarawa in Kamuela. The 4th Division was on Maui, where they had their own observation squadron, Marine Observation Squadron 4 (VMO-4). We worked back and forth on things. For example, we loaded out together but we really got the most of our training (earlier) at Camp Pendleton. During our time in Hawaii they told us about the Brodie Gear (landing system) and brought one there to show us how it worked.[2] While the commanding officers (C.O.s) of VMO-4 and VMO-5 were going to try using this system, the rest of us got on this LST with this gear on it. Our C.O.s were going to try to land with their aircraft that had hooks mounted on them (to use this landing system). The VMO-5 C.O. (First Lieutenant Roy Miller) tried first to land but he misjudged the distance and went into the drink but was rescued. Next, the VMO-4 C.O. came in trying to land his aircraft. He got the hook OK but either he was going too fast or there was too much tension on the cable above so this C.O. swung upwards and cut the cable with his prop. Therefore, the only two aircraft that we saw retrieved with the Brodie Gear went into the drink. However, when we went into Iwo, they found it was much easier to launch aircraft with this gear then to receive them. I recall that four of our planes came into Iwo after launching from their (specially equipped) LSTs. All of the others came in on aircraft carriers; two to each ship.

Original artwork for the official patch of Mert's unit, VMO-5. (Mert Hansen Collection)

While training in Hawaii, Mert was able to reconnect with younger brother Burnett (Barney) Hansen (left in the picture). They were able to meet because Mert knew that his sibling, who was a member of a naval underwater demolition team, was training in Maui and he found Barney by flying low over their camp! (Mert Hansen Collection)

Members of VMO-5 before leaving Hawaii, bound for the Pacific frontlines. (Back row, left to right): Jim Link, Leon Ellsworth, Jacque Benard, Tom McCawley and Clyde Fitzpatrick. (front row, left to right): Alvin Varner, Merton Hansen, Roy Miller and Jerry Lockwood. All men are second lieutenants, except Varmer and Miller, who were first lieutenants. Ellsworth was destined to never return home from combat. (Mert Hansen Collection)

*****

After his training was completed in Hawaii, on 22 January 1945, the young Marine aviator and his counterparts flew to the island of Guam via an Army C-54 transport. Once on Guam, they learned that the unit would be heading into their first combat on Iwo Jima. Next, their aircraft were reassembled and then flown to the islands of Saipan and Tinian. On 12 February 1945, while on Saipan, Mert and his squadron-mate Leon Ellsworth boarded the USS Sargent Bay *for their journey to the frontlines. According to Mert's recollections, two men and two aircraft (from his group) went on each carrier for the voyage to Iwo Jima.*

The Battle of Iwo Jima

Off of Saipan, our aircraft were taken on board with a derrick. (Prior to this), the captains of these carriers (at least ours) never saw a plane like mine, I think. Since this was done at sea, they had an awful time doing this loading. The man who was piloting this little ship that took us (Ellsworth and I) and my plane wasn't too skilled with operating his boat. He banged his boat into the side of the carrier so the captain was getting angry with him. I stood up on the gunnels (from the small boat) and I had to get on this carrier because my plane was on it. There must have been an eight or ten foot difference between the carrier deck and this small boat. So when my little boat went up (with the wave) and the carrier went down, I grabbed the edge of this ladder (on the side of the carrier). I was hanging by the tips of my fingers and then the carrier went up with me trying to hold on; it was probably fifty or sixty feet from the water. I didn't have any time to think when two sailors (one on each side of me) grabbed my wrists and flipped me on the deck. From above, the captain was watching all of this so I gave him a salute but he half-heartedly saluted back. I think back now on this and see what a stupid thing this was to do.

*****

The actual loading of two VMO-5 planes off of Saipan from a smaller ship (number 78 on the hull) to the USS Sargent Bay. *Mert and his squadron-mate Leon Ellsworth also were traveling with these two aircraft so this is the actual scene that Mert is describing. (USN via Duncan Cameron)*

While anchored off Iwo Jima, Mert and his buddy Leon Ellsworth stayed on board the Sargent Bay. *On 21 February 1945, a menacing surprise appeared on the horizon, it was a formation of* kamikaze *aircraft led by Lieutenant Hiroshi Murakawa. This group was called the 2nd* Mitate-tai *("Imperial Shield Unit") consisted of twelve* Suisei *carrier bombers, eight* Tenzan *carrier attack planes and twelve* Zero *carrier fighters escorted the group. Their target was the ships participating in the Iwo Jima operation. The escort carrier* Bismarck Sea *(CVE-94) which was moored alongside* Sargent Bay *was struck by two of these suicide aircraft. This successful air attack caused heavy damage and two large explosions engulfed the ship. As the spread through the now doomed ship, the magazines blew up, one after another. Then, a large explosion tore the* Bismarck Sea's *stern off and it sank, taking 318 men to their graves. Later, the carrier* Saratoga *also was struck by a total of seven of these* kamikaze *attack aircraft and five got successful hits. This lethal attack caused large fires in the hangar deck and major damage on this once grand ship. After* Saratoga's *crew got the carrier under control, extinguished the fires and recovered its aircraft, the ship evacuated the combat zone. Then, it was sent to Bremerton, Washington for repairs but didn't participate in the war effort until late May of that year.* Saratoga *had 123 men dead or missing after this engagement.*[3]

*****

When they called the alert, boy, Ellsworth and I got on our life jackets then went to bow section of the carrier because it seemed like the safest place to be with all of this metal around us. The carrier deck extended out, just above us. I think, I left a facial impression on that steel deck when it was all over because we laid flat. The *kamikazes* came in and we saw them hit the *Bismarck Sea*, which later sank.

*****

As mentioned in Ed Fox's earlier story D-Day for the invasion of Iwo Jima was on 19 February, but Mert didn't land on the island until the 28th, nine days after landing operations officially began. His unit, Marine Observation Squadron 5, and their companion group, Marine Observation Squadron 4, would be based on the same airfield, Motoyama One, on the southern part of Iwo Jima. However, each group would be based on opposite ends of the airstrip, in case one of the squadrons lost their observation capabilities due to combat. Remarkably, Motoyama One, was not really secure from enemy fire at all and these brave aviators had to conduct flying operations while constantly fired upon from the Japanese.

*****

After the *kamikaze* attack, I wanted to get off this carrier and away from them. But, it wasn't until 28 February that I could take off and fly to the island. I asked an operations officer what direction and distance Iwo was from our current position. He pointed his hand outward and said, "Go that way for about 100-150 miles." Then, I asked for the compass heading to take and he gave me one. Next, I spoke with the captain but he thought that I better not go and that my aircraft was loaded too heavily. Because of this, he felt that we would never take off (safely). I was flying with an observer and I told him that it could hold twice as much. The captain reluctantly gave me the go ahead. Then, I got back to my plane and once again the captain wanted to talk with me because he felt that we couldn't take off. I told him that the load was fine since we readjusted it. At that time, all I could think about was the *kamikaze* attack that happened on the previous night. He told me that we only had 300 feet of runway but I replied that this was just fine. Then, the captain said that I should go ahead and try. I then learned that the combined wind and the speed of the carrier together would be approximately forty-six knots. I told (the guy on deck), "This thing will fly with forty knots and I will be flying at 1,000 feet before I get to the bridge (of the ship)." He replied back, "Don't be foolish." Well, so help me, I got up to 1,000 feet and the carrier was still pulling away from me. Then, I flew past the captain's bridge and gave him a salute. He just waved his hands back at me.

When I got about ten minutes from the carrier and looked back, I couldn't see anything. This was the weirdest feeling and I hoped that I was going in the right direction. I saw nothing but water and this was an awful feeling. Well, after flying for about an hour, I looked up and saw two F4Fs (Wildcat fighters) escorting us. Ellsworth was on the same carrier and took off behind me. I flew in and landed on the island. The runway that they had was 300 feet. On the left side when you were coming in from the east was Mt. Suribachi and all of the troops were ours. On the north side, the Japanese held that area while the south side was ours. So, you couldn't come in from the west and the wind was from the east. When I was landing, I made a ground loop and headed to the east as fast as I could. But, I didn't realize that the Japs held so much land on the other side of the runway so close to us. When I taxied down to the end, I saw a guy who was standing and waving at me from a foxhole. So I pulled up to where he was and it was my mechanic. He had a message that the 13th Marines wanted my plane in the air to do some spotting. I said that I should gas up and asked where the fuel was. He said that they rolled in a barrel yesterday so we could gas it up. When the mechanic went to gas up the aircraft, he got shot, so, I had to gas it up myself. In order to do this and reach the tank on top of the wing, you had to put one foot on the wheel and the other foot on the strut. But, when you did this, you made a nice target. I got it gassed up and took off, but don't remember anything about that first flight.

The thing was that the operation was so darn small and there wasn't any room for the artillery to get ahead. Therefore, for a week or ten days, we would land and take off under the projectiles from our artillery; they were shooting over us. Normally, we would be behind the artillery, between the headquarters and artillery, so you could see it from there. You would head out to sea when you took off then get some altitude until you got oriented and "see" everything. I don't know if my buddy, Second Lieutenant Leon Ellsworth got shot down and burned from our own trajectory or from Japanese fire. However, I did see a Navy observation floatplane up there that came from one of our cruisers. I also could see the trajectories coming from our ships. I could see that they were just missing him up there but we were on a different frequency. I was trying to call the headquarters of the 13th Marines to call that ship so they could tell that pilot to fly higher. While we were trying to get a connection, I think, one of our cruisers hit the plane and it was gone.

*****

On 13 March 1945, Second Lieutenant Leon W. Ellsworth was killed in action on Iwo Jima. He was from Ellensburg, Washington.[4] *The young USMC aviator flew twenty-six missions over the island before being killed. Posthumously, Ellsworth was awarded the Air Medal with two Oak Leaf Clusters, the Purple Heart and the Distinguished Flying Cross.*[5]

Mert Hansen (right) and an unidentified mechanic on Iwo Jima. They are pictured with Stinson OY-1 (U.S. Army Air Force serial number 42-99512, later redesignated USN/USMC Bureau number 02747). Just a few years ago, this historic aircraft was restored to its original wartime condition by Duncan Cameron. (Mert Hansen via Duncan Cameron)

Motoyama Airstrip Number One on Iwo Jima was where both VMO-5 and its counterpart, VMO-4, operated from. This perspective was taken on 5 March 1945 and is facing south towards Mt. Suribachi. (USMC via Duncan Cameron)

Ground crew with Second Lieutenant Leon Ellsworth's aircraft "Yankee Gal" just before both were lost in combat over Iwo Jima. (Martha Link via Duncan Cameron)

Combat Missions over Iwo

We were connected directly to the 13th Marines, the same as a forward observer would do. Typically, the men that flew with us were forward observers. They had to learn to estimate distances from the air, instead of the ground, which could be hard to do. We flew between 25-100 feet above the ground for our missions. My Distinguished Flying Cross wasn't for a "specific" mission. It was awarded for knocking out some enemy gun emplacements. I flew thirty-eight missions and they were short ones because you could see a lot of a 3x6 mile island in a brief time. I flew with a different observer on each mission; they were in the 13th Marines though. Sometimes, they would get hit in their legs, which were exposed. One time, an observer yelled out that something had hit him. Luckily, the shell struck his map case and it only went a quarter through it, otherwise he would have been wounded in his chest.

One day I was out there and you could see quite a bit. However, you had to get down low to see people (below). As we were coming in from the south, there was a hill with some cliffs descending behind it. Right over the island there was a 100 foot ceiling but it was real funny because you could go out and see five-six miles and there were no clouds at all. I spotted these guys jumping down and hiding behind these cliffs. We marked the section that they were in and then our artillery started shooting in there. When we came back to this area to see what happened, we noticed that very few of them were hit. This is because they were standing real close and tight against this cliff but we couldn't bring any naval fire in because if you were a few feet too high, they would hit our lines. When I flew through there they waved at me and I waved back. But, after they did this the third time, I got really disgusted so I took out my .45 pistol and shot it a few times. This was stupid though because I didn't want them to think we had any guns. Since I knew this wasn't working, I called in a "time fire," when they would send in artillery shells that would explode twenty-five to fifty feet off the ground. So we ordered that and got out of the area. When we came back, there were bodies all over the place. At that time, I thought it was great but when I think back now, it was just awful. We made a lot of families sad that day.

On another occasion, I saw them (Japanese troops) roll a hidden gun out of a chute. So, I radioed (this target) to a flight of dive-bombers that were in the air. They came in a few times but were missing it by a mile. I said, "Watch me now." Then, I flew about twenty-five feet above this gun, dropped my wing and made a quick turn above it. "There now take her. Did you see where I turned?" Then, the C.O. (of the dive-bomber unit) replied, "Is that you, Swede?" I asked him who he was, and it was the C.O. from the carrier that I was on. He then radioed to his men, "That is Swede down there; you saw what he did with his plane to show us where that target is, we have to take it out for him." Then, came over the air were a bunch of okays from the other pilots. They came screaming downward and they let loose at about 1,000 feet; I thought they would kill themselves. The SB2Cs (Helldivers) that they were flying had a habit of losing their tails in a dive. They took this target out and attacked it from all directions. The Japanese used a railroad track to move this gun in and out of the hole. This occurred a few weeks after Mt. Suribachi was secured.

One of the most gruesome things that I saw occurred one day on a plateau area on the island. Two of our guys were chasing this naked enemy soldier. All of a sudden a third man (one of ours) opened on this Japanese soldier and he was engulfed in flames then jumped off this high cliff. At that time, I thought it was funny but if I saw it today, I would have shot the man with the flamethrower. During my time on the island, I only saw one enemy prisoner and he was a graduate of Cornell University. This enemy soldier looked up at Mt. Suribachi and said, "Oh, you got all that stuff up there already. We have been working for years to get something up there and now after only two weeks you have done it. He was very sick and threw up this long, dark green worm. Since I hadn't eaten anything for about a week, I was fine, otherwise, I would have.

When the Army came in they set up directional landing lights but kept on giving me a red light so I had to circle around again and again. But, I really had to go to the can so I turned my switch off, waggled my wings hastily and landed under the wing of a B-29, rolled down the runway then restarted it again and took off. You could run that aircraft like a glider and it would glide for a long ways. It was the nicest aircraft that I ever flew. This aircraft could get you out of so much trouble. You could do stupid things but if you woke up soon enough, you could get out of trouble easily.

The stark landscape of Iwo Jima as seen from the air from Mert Hansen's aircraft. (Mert Hansen Collection)

Another time (while I was flying at night), my I.F.F. went off on my plane and every time I tried to land, the ships started firing at me.[6] I told them who I was but they wouldn't listen to me. The Marines couldn't get this message to our ships. Finally, they kept me so long up there, I told them that I was coming in at water level and they would hurt more than me if they kept shooting. I got around to the end of the airport where I thought this runway ended. It was dark but there were shadows and you could see a bit. My friend Fitz had drove a jeep down at the end of the runway. He said, "Swede, watch for the flash, watch for the flash." I replied that I was watching for it and he quit talking. I got around and "bing," there was a flash. Sure enough, I landed and Fitz was there with a jeep.

Our missions were flown "as needed." For example, a daytime search to see where this mortar fire was coming from. This happened when I first arrived on the island. We would check each quadrant on a map and try to mark the area that the mortar fire was coming from. It was possible to communicate with both the troops on the ground and our ships. However, the latter (ships) had to know you were coming so we could use the same frequency on the radio. I forgot how this was done; possibly the artillery people would notify a ship that we wanted to talk with them. Going into the battle, I felt pretty well-equipped. When we were at Quantico, I had to do some ground firing before doing aerial gunnery so I knew a little about this beforehand. I also could tell quite a bit of difference between the aerial observers that I flew with too. Some of them had much better judgment estimating the distance between points, etc.

Coming Home from Iwo

The Japanese had built these big cisterns along the runway (out of rock). They were about twenty feet wide and about forty or fifty feet long, with a depth of seven or eight feet. When we got ashore, we filled one end up with sand, barrels, etc., but left a hole in the side. On top of this, we put metal strips, canvas and maybe four-six inches of sand. We knew that people could run across the top of it and wouldn't realize it. Four of us slept in it on the first night but then two of the men moved to another one where you could stand up. Ellsworth and I stayed in this hole together but as I mentioned before he was later KIA on the island. One time, we heard these Japanese troops just outside of where we slept. Ellsworth and I sat on each side of the entrance area with our knives but they just walked right past our hole. I said that we should let the guards take care of them.

I was on Iwo for about thirty days and it was long enough to lose thirty pounds. There was plenty of food but I never took the time to fix it. The food was in cans and they had this one little place where you could go and pick what you wanted. There was a barrel, on blocks, over a fire so it was possible to warm up your food. We (pilots) were on good terms with the medical department and they gave each of us a little box of brandy. In the morning, we weren't able to brush our teeth very good so we used it as mouthwash. Of course, some of it got swallowed too. This also kept your mouth and throat clean.

You could see artillery being fired all night long and these shells were going overhead (our position). However, it took a few rounds of artillery to know if they were coming or going. We also tried to keep a survey plane up at night. One night when I was flying over the island, I got very disgusted. I went up higher and we could see more flashes below. They shot up this parachute with a big white flaming object underneath it; this illuminated the whole island. This was shot in the air just above me and I thought, son of gun, now I am a perfect target!

As an anecdote, everyone in my group were second lieutenants and we left Iwo Jima, most of us were first lieutenants, except for maybe three or four guys, I think. But, several of us who earned promotions could not even pass a physical exam (after the battle). The doctor said, "What in the world have you guys being doing?" We all failed both the eye and hearing tests. One of the guys replied, "We were bumming around in the sand on Iwo Jima." Then, we were ordered to go a rest home on Oahu for about a week.

Mert and his buddy Second Lieutenant Leon Ellsworth lived in these quarters while stationed on Iwo Jima. (Mert Hansen Collection)

Mission planning at VMO-5's command post on Iwo Jima. (Martha Link via Duncan Cameron)

Members of VMO-5 with a captured Japanese flag just outside of Mert's living quarters. (Left to right): Jerry Lockwood, Tom McCawley, Jacque Benard, Roy Miller, Merton Hansen and Jim Link. At that time, all of them are second lieutenants except Miller who was a first lieutenant. (Mert Hansen Collection)

Per his logbook, Mert's last combat mission was flown on 16 March 1945. During his time on Iwo Jima, he flew thirty-eight missions. However, his logbook only shows thirty-seven entries and does not contain one for 28 February, the day he took off the carrier and landed on the island. However, the veteran aviator knows that he went up shortly after landing that day but after his mechanic was hit by sniper fire, Mert may have forgot to write down this mission in his logbook. The only combat fatality for VMO-5 was Second Lieutenant Leon Ellsworth. VMO-4 lost several men in combat but none on Iwo Jima. They also participated in missions on Saipan and Tinian.[7]

On 25 March, this Marine boarded the USS Sea Sturgeon *and two days later, he left Iwo Jima bound for Hawaii. Mert arrived safely in Hawaii on the 14 April and was fortunate to survive the battle without any physical wounds. Nevertheless, his wartime experiences clearly affected how the man lived his postwar years until today. In late December 1945, he was relieved from active duty and stayed in the Marine Corps Reserve until retiring on 1 December 1958 with the rank of captain.*

As a related anecdote, Mert recalled the terrible Japanese suicide attack that occurred the day after he left Iwo Jima. In the early morning hours (around 0515) of 26 March 1945, about two or three hundred Japanese troops came swarming down from the northern part of the island (along the west side). Their intent was not a suicide charge; instead, they focused their attack on the Marine and Army bivouacs which included Motoyama One where Mert's unit was based. The area the well-armed enemy attacked was nearby the western beaches and lasted for about three hours. The Army aviators (from the VII Fighter Command units were overwhelmed during this surprise attack but were able to regroup and fight back. Marines from the 5th Pioneer Battalion (5th Marine Division) quickly formed up their lines and were able to stop this unexpected attack. As a result of this combat, the Marines had nine killed and thirty-one wounded while the Army airmen lost forty-four men along with eighty-eight others wounded.[8] *The Japanese side lost 196 men in this last-ditch combat.*

Sadly, some of the casualties of the American side could have been averted if they heeded the earlier advice from their Marine counterparts who were getting ready to leave Iwo Jima. Just prior to this terrible attack, the Army pilots who landed on the airfield declined to live underground, like the Marines had done. Instead, they lined up their tents along the runway and therefore, the Army airmen suffered terrible casualties that day. According to Mert, their counterparts in the Army seemed to have a disdain for their fellow Marines because they looked so disheveled and had lived underground. He said his commanding officer, First Lieutenant Roy Miller, and executive officer, First Lieutenant Alvin Varner, had met with the Army pilot's leadership and they had advised the Army pilots to "dig in." The Army pilots' reaction was that they weren't afraid to set up their tents above ground, inferring the Marine pilots were cowards for being underground. The Marines tried to explain that official announcement about the island being secured was Marine speak meaning their position was secure and there was now no chance that they would not prevail However, this did not mean it was not still dangerous.[9]

Postwar

On 26 August 1945, Mert Hansen married his sweetheart Jeanne Mack. Together, they had two children, Lona and Stephen. The couple also has two grandchildren, Erin and Joshua. In January 1946, Mert enrolled at the University of Minnesota and later graduated with a degree in Ag Education. After graduation, he went to work for Hy-Line Poultry Farms of Johnston, Iowa. Mert worked there as a geneticist for the next thirty years. When Hy-Line was sold to a German firm, he decided that it was time to retire. By this time, Hy-Line had about a million chickens which were on contract with various farmers throughout the Midwest. One of the board members had bought the flocks and asked Mert to manage them. Once this project was completed, he took over management of a real estate project in Norwalk, Iowa but, at the age of seventy-five, Hansen finally retired. When I was visiting his home, the ex-Marine recalled an incident that occurred a few years after the war when he just returned to work after a vacation.

*****

HENRY WALLACE, WHO WAS the head of the poultry division, had sent a Japanese boy to the farm where I worked as a geneticist. After being on vacation for a few weeks, I thought that I should go to the range and check on my birds. There, I spotted this Japanese boy standing there with his arm stuck in the gate. His fingers were caught in the wire that kept this gate tightly closed. He had tried to release the wire but he couldn't get his hands up (to get it off) so the boy was stuck there. I knew right away John did this. He was an ornery guy but had a good heart. I went there right away to release this Japanese boy. Three years earlier I would have laughed like a son of a gun, but that day, I felt really sorry for him. John said that he thought I would "like this." My reply to him was, "He is a human being."

The reason why he was sent over was because his father was interested in selling our chicken over in Japan. I eventually met his brother too and both of them had grown up in this country during the war. Their dad was an agriculture attaché but couldn't go back to Japan when the war broke out. He also bought some boars here and brought them back to Japan.

*****

MERT'S SON STEPHEN shared a very interesting story that happened during his childhood regarding his dad. Stephen and his sister Lona were instructed that one drawer in their parent's bedroom dresser was "off limits" and it could not be opened for any reason. Of course, one day, the kids wanted to see what "secret" was in this drawer. So they opened it and the drawer was full of their father's military items such as Distinguished Flying Cross, USMC insignia, etc. Mert wasn't too pleased that day, but in the end, everything was put in a frame where it still is today.

During his life after the war, Mert continued to stay busy in his free time including serving as a 4-H leader, Boy Scout leader, Little League baseball coach and Sunday school teacher. As mentioned earlier, he also was a charter member of St. Paul Presbyterian Church, where he oversaw its construction and served as an Elder. This veteran served as a member of the City Council and was a member of the local school board. In 1986, he was named Citizen of the Year by the Johnston Chamber of Commerce. Later in his adult life, Mert's son, Stephen once asked his dad with some frustration, "Why do you stay so busy?" In a telling moment, the old Marine said that he sometimes stayed up a night asking God, "Why did I live and so many others didn't?" So, this was his way of "making this right" after surviving one of the costliest battles of the Second World War.

Reunited with an Old Friend

SOMETIMES, REALITY CAN be more amazing than fiction. Fast forward to almost seventy years since World War II ended and a surprise phone call come into the Hansen residence in Iowa. On the other end of the phone, was Duncan Cameron who found Mert's name while researching about his squadron, VMO-5, at the National Archives. Duncan's first question was whether Merton lived here? Jeanne Hansen replied "Yes, he does." Then, Duncan asked, "Was he a Marine aviator in World War 11?" She replied, "Yes, he was." After this, Duncan said that he was restoring an aircraft that Mert may have flown on Iwo. Of course, the Hansens were very surprised and interested. Remarkably, then Jeanne Hansen pulled out her husband's pilot's log book and Mert asked Duncan, "Which Bureau number?" "02747" was the reply back from the caller. Then, the Hansens scanned through Mert's log book; they found numerous entries from the Battle of Iwo Jima and confirmation that he flew it in combat. Just incredible. Prior to this, Duncan knew that this aircraft was shipped to Hawaii with VMO-5 and was assigned to the squadron during the time the battle took place, but he didn't have any confirmation if it had flown combat missions on Iwo Jima. Next, Duncan visited the Hansen family and he interviewed Mert. During this meeting, Mert shared a picture he had of himself and 02747 after the battle. This changed everything for both he restorer and the former original pilot of this rare warbird. I will let Duncan continue the story.

I had started working on the plane prior, but now knowing what I knew, and that Mert was alive to be united with his plane, I started (with the blessing of my wife) a "full court press" on 47. I worked very literally sixteen hours a day everyday that I was off work to get her done while Mert was still around to enjoy her. This was a responsibility I took very seriously because so few pilots have ever been reunited with the actual plane they flew (not a replica). This was my second OY restoration; my first having won its Oshkosh class in 2008.[10] I was about sixty-seventy percent done when Tom Westfall, having left another restoration facility that was going under, stopped by to visit in Tennessee on his way home. I hired him to help with the fabric covering. As time went on and deadlines approached I asked if he would help me full time, splitting shifts as we, toward the end, worked seven days a week and sometimes, twenty hours a day preparing for Oshkosh. It wasn't ready and subsequently after I applied to National Aviation Heritage Invitational (held at the annual Reno Air Races) toward that goal. Tom worked with me side by side for the remainder of the project.

*****

Duncan and Tom meticulously restored this OY-1 to its 1945 condition during the time that Mert flew it, done to the rudimentary patches to fix the combat damage on the airframe fabric. Even more amazing was that they, not only completed this project in less than two years of hard work, but also achieved their goal of having Mert fly in his old wartime aircraft. It was a life-changing event for everyone involved and Mert's wife, Jeanne, said that this airplane

Mert's former Iwo Jima-used OY-1 is beautifully restored, and now flies with the Texas Flying Legend Museum. (Moose Peterson/Texas Flying Legends)

and what Duncan and Tom did for them added years to their lives. Their hard work also was noted by the judges at the 2012 Reno Air Races, when they were awarded the prestigious Neil A. Armstrong Aviation Heritage Trophy for their hard work and attention to detail to make everything just right. Later, this rare combat aircraft was sold to the Texas Flying Legends (TFL) Museum and it flies with several other rare World War II era warbirds as a flying memorial to these heroic aviators that flew dangerous combat missions so many years ago and without much fanfare. Duncan is still the chief pilot for this rare OY-1 and flies it on behalf of the TFL Museum when they participate in air shows or other events.

Almost seven decades passed since the last time Mert Hansen flew in "his" aircraft. This happened due to the hard work and perseverance of Duncan Cameron. (Duncan Cameron)

The author with Mert Hansen during his visit with this incredible but very humble veteran. (Ron Werneth)

Reflections

Of course, I was pissed off; especially when you had your best friend knocked off. Laying there at night, I thought, "Why the hell did they have to do that?" I don't think I felt this way particularly towards the "Japanese." Yes, it was "a battle." I would have killed any one of them if I thought they had a chance of killing me. In retrospect, just after that mission where I saw so many Japanese troops killed, I felt pleased. But, later on in life, this was one of my first regrets.

[1] On 5 October 1945, Fleet Admiral Chester W. Nimitz , the supreme commander of the allied naval forces was honored with a parade and speech in Washington D.C. to celebrate the ending of World War II. One thousand aircraft took part in a flight above the parade in his honor.

[2] The Brodie System was an amazing yet complex method where a cable system mounted alongside of a ship allowed it to do aircraft take offs and landings, without any deck. This system enabled L-5s to operate under very remote conditions so they could be utilized in battle for combat missions for aerial spotting and reconnaissance.

[3] Ron Werneth, *Beyond Pearl Harbor*, 82.

[4] USMC Casualty Card for Leon W. Ellsworth.

[5] C. James Quann, *WSU Military Veterans: Heroes and Legends*, 39.

[6] I.F.F. stands for identification, friend or foe. Unfortunately, it could only be used to identify "friendly" targets but not ones that be operated by the enemy. Furthermore, if no reply or an invalid response came from the other target, it was hard to determine if it was friendly or a possible opponent.

[7] R. Duncan Cameron, *Consolidated Stinson OY-1 BuNo 02747*, 11.

[8] *C-3 Periodic Rpt* No. 35, 26Mar45; *5th MarDiv SAR*, Annex W, 2.

[9] This story about the Army aviators and the tragic Japanese attack on 26 March 1945 was shared to me by Mert's son Steve, who spoke to him at great length about it.

[10] Duncan is talking about the world-famous, large annual AirVenture air show held by the Experimental Aviation Association (EAA) every summer in Oshkosh, Wisconsin.

Chapter 9

"We were going to kill them until they would sit down with us and talk peace."

Captain Grant C. Young (USN)

I MET CAPTAIN YOUNG, a local Navy veteran, during a "roll-out" party for a freshly restored SBD Dauntless dive-bomber now on display at Midway Airport in Chicago. Young is a distinguished naval airman whose career spanned from World War II until the Cold War. Throughout his nearly three-decades of military service, he participated in numerous combat missions and has many "highlights;" from flying with a youthful pilot named George H.W. Bush to putting a torpedo into the mighty Japanese battleship Yamato. *During my visits with Captain Young, the thing that was very clear to me was the "love" that I saw between he and his wife Ethel, even after all these years of being married. She softly calls him "Jack," his long-time nickname after his grandfather. Together, they had three daughters and one son.*

Early Years

GRANT C. YOUNG WAS born in Dixon, Illinois on 8 November 1921 and raised on a farm in Lee County (near Nachusa, IL.). He graduated from a one-room country school where the lad became good friends with a neighbor, Ethel Shippert. After graduating from high school, Grant worked for a metal fabrication shop in Dixon and helped his father with farming from June 1939 until August 1940. The eager eighteen year-old joined the United States Navy in August of that same year and did his basic "Boot" training at USNTC, Great Lakes, IL. Grant was assigned and attended the aviation metal smith school at Naval Air Station Pensacola, Fl. and served as an enlisted man for two years. Around the time Japan attacked Pearl Harbor, and the United States formally entered the war, Grant had taken and passed the entrance exam for OCS (Officer's Candidate School). He was ordered there in November of 1942 but having his heart set on

Grant in his Navy dress whites. This wartime picture was one of his wife Ethel's favorites. (Grant Young Collection)

being a Navy pilot, opted to attend Pre-Flight School instead, and following that, entered flight training. He was subsequently commissioned an Ensign, USNR, and designated a naval aviator on 5 October 1943.

Teacher Miss Bennet's 5th Grade Class picture from one-room Dysart School (South Nachusa, IL.) during the first week of September 1930. Grant is in the front row, second from the right, while his future sweetheart and later wife, Ethel Shippert, is standing third from the right, wearing a white dress and round-rimmed glasses.
(Grant Young Collection)

A young, dapper Grant Young dressed up in a suit.
(Grant Young Collection)

*****

I ENLISTED IN THE Navy in 1940 because of the opportunities it presented and I knew they had a program for enlisted men to go through flight school and you didn't have to have a college degree. Plus anyone with half of an eye could see that war was looming because of the German victories in Europe. Sooner or later we (the United States) would be drawn in. I don't know if anyone else did but I envisioned coming into a conflict with Japan. However, I didn't enlist with that "purpose" in mind.

One of my early childhood heroes, Howard W. Crews, went out to be a Navy pilot and got his wings in 1938. He was the backdoor neighbor to my grandparents' house, which is where I grew up in Dixon, Illinois. I lived with my grandparents because my mother died when I was thirteen months old. I lived with them until I was eight years when my father remarried. My uncle lived in our house and he was good friends with Crews. So, I got to know him very well. I just wanted to be a Navy pilot and everything that I did was towards that end. From the age of five years old, I was drawn to airplanes. The local National Guard held their annual maneuvers at Camp Grant in Rockford, Illinois and my grandfather held me up to look into a cockpit of an aircraft. I (thought), "Boy, I am going to drive one of these things if it is the last thing that I do." From then on, that determination is what kept driving me on.

Howard "Sandy" W. Crews was born in Dixon, Illinois, on 11 May 1914. Veteran historian John Lundstrom befriended Crews many years after World War II, in the late 1980s. He described him as tall and soft-spoken. Lundstrom said that veteran aviator was a terrific guy and very much respected. He shared with me some of Crew's background. Howard W. Crews joined the United States Navy as an aviation cadet in September 1937. In August of the following year he was commissioned as an ensign in the Navy Reserve and joined VS-42 on the Ranger. *Next, in September 1940, Crews went to Naval Air Station Pensacola and trained in aerial photography. The young airman was commissioned during March 1941, as an ensign, in the regular Navy. Then, in January 1942, he was promoted to lieutenant (junior grade). Crews joined VF-5 in March of that same year then later saw air combat while serving with his squadron at Guadalcanal. Because of his actions during air combat over Guadalcanal and destroying two enemy bombers, he was awarded the Silver Star. During the Pacific War, Crews led a VF-6 detachment on board the carrier* Princeton *(CVL-23). As an interesting anecdote, later in Grant Young's military career, during the Korean War, he got to serve with "Sandy" Crews in Reconnaissance Squadron 61 (VC-61). At that time, Crews was the commanding officer of that squadron. In his postwar years, Howard Crews stayed in the United States Navy until July 1963 when he retired at the rank of captain. He died on 28 January 1992 in Florida.*

VF-6 Detachment led by Lt. Howard W. Crews on board USS Princeton *Sept. 1943; kill markings are (1)* Emily *shot down 1 Sept 1943 by Richards L. Loesch off Howland Island. (2)* Emily *shot down 3 Sept 1943 by Thad T. Coleman. (*Emily *was the Allied code name for the Kawanishi Type 2 flying-boat). Left to Right, Standing: Charles A. Palmer, Joe D. Robbins, Richard B. Trimble, Ashton A. Roberts, Thad T. Coleman and Harvey C. Odenbrett. Kneeling: Edward J. Philippe, Howard W. Crews, Albert W. Nyquist, William A. Davis, Richards L. Loesch, Lindley W. Godson and John P. Altemus. (Captain Howard W. Crews, USN (Ret.). via John Lundstrom)*

Ditching in Lake Michigan

Ensign Young did his aircraft carrier qualification landings aboard the USS Wolverine *(IX-64), which was converted from a pre-war paddle-wheel steamer into an aircraft carrier. During the war, it and another ship, the USS* Sable *(IX-81) operated on Lake Michigan and were utilized to train thousands of naval airmen. However, there were sometimes training accidents and this young pilot had his first experience of ditching his aircraft in water.*[1]

*****

On 16 December 1943, right after my fifth carrier landing on the *Wolverine*, the engine seized and I landed in the water. There were little patches of ice floating around and I said, "No big deal." I thought that my watch wasn't even going to get wet. I went out on the wing and pulled the lanyard on my big four-man life raft. It went "poof."

I looked aft of the carrier and this rescue Coast Guard boat coming. He roared up and then slammed his diesel engine in reverse. I was on the knife-edge of the wing, trying to keep my watch dry. My aircraft is sinking and this sailor tries several times to throw me life-ring, but falls down. He said that the skipper could not get the engines started and I would have to get to them. By now, I am in almost knee deep and so I plunge out as far as I could. I inflated my Mae West (life jacket) and started swimming towards the boat. I began swimming as hard as I could. About half way to the boat I realized that I couldn't feel my legs anymore. I kept plowing around with my arms. My wedding invitation had just been mailed out for a wedding to take place on 27 December. I knew this girl since I was seven years old and we were going to get married. I got to the side of the boat and sailor said, "Come on, get out of there, it is cold!" I wanted to talk to him but I couldn't. They pulled me up, put me in a cabin and took my clothes off. I could move my eyes and that's all I could move. I heard this guy telling the chief what a lousy boat handler he was. I looked over and saw this guy was a lieutenant commander and had doctor insignia on. He said, "Look at that man chief, he is going to die and it's your fault!" I was thinking, "Do something you SOB, I am going to get married next week."

We got married and got on the Super Chief because I had orders to a squadron in San Diego.[2] On the second day out, I had the chicken pox. I had them on the beginning of our honeymoon. They met me with an ambulance when I arrived there. So, I didn't get to see my bride until about two weeks later. We had a brief three days together then I went down to San Diego to check into my squadron. But, they just moved to Seattle. Then, I was thrown into a pool of pilots at FASRON (San Diego). A week later I was on my way to the war zone on a transport type jeep carrier. The squadron that I was initially assigned to never went to war. They were assigned to a jeep carrier off the Peugeot Sound doing anti-submarine patrols along our Pacific Ocean frontier for the rest of the war.

Grant and Ethel's wedding photo. (Grant Young Collection)

*****

While Grant served overseas his wife Ethel worked as a coding/decoding specialist with the Federal Manhattan Project, makers of the atomic bombs. In early April 1944, the rookie Navy aviator Grant Young joined his first "front-line" unit, VT-10 (the Buzzard Brigade); while it was on board the carrier Enterprise *(CV-6). He flew a brand-new*

Avenger from the jeep carrier Barnes *(CVE-12) to the ship. At that time, the carrier was anchored off of Majuro in the recently captured Marshall Islands.*

USS Enterprise *(CV-6) in action, around the time that Grant joined this aircraft carrier. (USN via NA)*

1 March 1944, Majuro Atoll, part of the Marshall Island chain. This is where Grant joined the Enterprise while the atoll was being used as an advance base during VT-10's second wartime cruise. Note the men enjoying the beach. (USN via Joseph Houston and Steve Moore)

Grant in flying gear during this time period that VT-10 served on the Enterprise. *(Grant Young Collection)*

April 1944, on board the Enterprise. *In the center is VT-10 skipper Lieutenant Commander William Martin and his Avenger crew, left, radioman ARM1c Jerry Williams, and to the right of Martin is his turret gunner AOM2c Wesley Hargove. Sadly, on 13 June 1944, both Williams and Hargrove perished when their aircraft was shot down over Saipan. However, Lieutenant Commander Martin suffered just a bruised hip after the TBM crashed and was later rescued by a Seagull floatplane. VT-10's skipper eventually made his way back to the carrier and resumed his duties. (USN, via W.H. Belden Jr. and Steve Moore)*

Off to Combat

I JOINED TORPEDO Squadron 10 on the venerable old *Enterprise*. I caught them after they already had been out to sea for two months. So, I had a short six month cruise to finish off with them. I learned how to really fly off aircraft carriers and had a great skipper. They had already hit Palau and Yap islands before I joined VT-10. Our group did the first night bombing attack off an aircraft carrier under the leadership of Commander W.I. Martin. He hit upon the idea of using radar and attacking at night. They sent out twelve planes on a strike in February 1944 and lost one of them on the mission against Truk Island. They didn't know what happened to the aviators who flew this aircraft. I was a replacement pilot for one of the lost airmen.

*****

AS YOUNG MENTIONED above, Lieutenant Commander William Inman Martin was the commander of Torpedo Ten during the second cruise on the Enterprise. *On 1 July 1943, he took command of the squadron and led the unit for the duration of the cruise. Martin is a well-known pioneer of flying at night and using radar. Unfortunately, he could not personally lead his group on the first night attack mission where they would utilize radar and fly at low-altitude. This is because Martin had previously suffered a broken arm and was grounded until recovering from this injury. Nevertheless, he was the man who always pushed VT-10's pilots to be proficient in this type of flying.*

The squadron had trained hard to prepare for this bold night attack which occurred on 17 February 1944. This mission was one of several flown against the Japanese stronghold on Truk Atoll on the 16-17 February by the Navy airmen in Task Force 58. They were trying to neutralize both Japanese shipping and land-based installations on Truk Atoll. However, it was a lot more dangerous than the ones done in the daylight hours. The "target" for this respective night sortie would be the enemy's shipping in Truk Lagoon. Twelve TBFs took part in this mission with the three division leaders being the veteran pilots, Lieutenant Van V. Eason Jr., Lieutenant Russell F. Kippen and Lieutenant Robert S. Nelson. Take off was around 0400, early in the day, on the 17th. Armament for each aircraft was four 500-pound bombs with 5-second fuses. Everything went as expected that night except for the tragic loss of one TBF which was flown by pilot Lieutenant (junior grade) Lloyd Nicholas and his crew of ARM3c Wilbert W. Dockter and AMM3c Len C. Thorton. In summary, VT-10 had thirteen direct hits and seven near-misses for this daring mission. Ensign Young joined Torpedo Ten about two months after these missions against Truk Atoll.[3]

*****

AS THE NAVAL AVIATOR recalls below, Young did not participate with VT-10 in their air operations in the Marianas, but was able to get some valuable flying time as a temporary member of VT-51.

Flying with George H.W. Bush

JUST BEFORE THE Marianas Campaign, three TBFs from the USS *San Jacinto* (CVL-30) were lost and nobody had an idea on what happened to them. There were a lot of anti-submarine patrols to do and I was sent to that ship. I spent three weeks as a shipmate of (future) President George H. W. Bush on board the *San Jacinto*. George was just another pilot and I had more flying time than him. It was a light carrier in our task force. They lost three aircraft off Majuro

Atoll (in the Marshall Islands). A couple of days later an admiral decided that one pilot and aircraft (but no air crew) from each of the other three carriers should go to the *San Jacinto*. The transferred men and their aircraft would stay there until we met up with one of the jeep carriers and got more planes. I (Ensign Young) was the junior officer from VT-10 so I went. I spent three weeks flying with Bush and flew two hops every day. These missions were either anti-submarine patrols or attacks on one of those islands. I wore the same flight suit for three weeks and temperature was around 90F every day. George H.W. Bush actually flew me back to the *Enterprise* when it was time to go back. He was later shot down on a mission (after I left).[4]

Future United States President George H.W. Bush as a young, naval aviator in his TBM Avenger during World War II. (NA)

*****

Young returned to the Enterprise *on 24 June 1944 at 1458.*[5] *In August 1944, they finished their tour because VT-10 had been out there for three months before he joined. So, this aviator went to the states (east coast) with the squadron to train.*

1945 – Grant's Second Cruise

During this time period, I joined the carrier *Intrepid*. In February 1945, we deployed from the states and went to Hawaii for a few days. We got to fly a few training missions and then had an operational readiness inspection by the admiral's staff. They towed a spar behind the ship and we bombed it and the fighters strafed (the target). By then, we were a well-trained air group. Our air group commander (CAG) was Lieutenant Commander John Hyland. He was Admiral King's pilot when he was in Washington. He was a good administrator, hell of a pilot and up in the air as much as we were.[6]

The aircraft carrier USS Intrepid *during the fall of 1944, just a few months prior to Grant going on board with VT-10 for their third wartime cruise. (USN)*

Commissioning picture of the VT-10 taken on 15 September 1944 at Naval Air Station Quonset Point, Rhode Island. These sixteen officers (and forty-two enlisted men) would make up the squadron for its third wartime cruise. Front row kneeling, left to right: Lieutenant Commander John Lawrence (CO), Anthony Rasmowich, Edward Laster and Sidney Overall (all ensigns). Middle row, standing: Grant Young, Leo McNamee, Orbie Guthrie, Bailey Badgley, Noble Atkins and Willard Miller (all of them are ensigns, except Atkins, who is a lieutenant). Back row: Arthur Fulton, Richard Kimbel, Robert Paine, Charles (Gene) Hudson, Harry Hunt Jr. and Charles (Charlie) Hudson (they are all ensigns). (Naval National Historical Foundation, Operational Archive Branch, VT-10 Aviation History via Steve Moore)

The senior officers for Air Group Ten during this cruise. Left to right: Wally Clarke (Fighting Ten), Will Rawie (Fighting-Bombing Ten), John Hyland (Air Group Commander), Bob Buchan (Bombing Ten) and Larry Lawrence (Torpedo Ten). All of these men are lieutenant commanders. (USN photo, via Arlo E. Schroeder and Steve Moore)

The Battle of Okinawa

1 April 1945 was the day that Okinawa would finally be invaded after several days of air strikes to help "soften up" the last Japanese-held territory before a possible invasion of the mainland of Japan. Eleven of VT-10's aircraft were scheduled to fly (in support) of the friendly forces doing their landings. One of these VT-10 Avengers was flown by Lieutenant (junior grade) Grant C. Young, AOM2c John Floyd (turret gunner) and ARM2c Adolf Kolodzik, but he went by "Bill" Kolodzik. Initially, the airmen were disappointed to learn that they would not be directly supporting the landing troops. Instead, the group would be assigned targets about a mile or so away from the invasion landing zones. In the end, however, this point was a good thing because VT-10 would have a greater variety of targets ranging from enemy tanks to land fortifications.

The pilots were ordered to man their planes at 0710 and thirty-five minutes later all of the aircraft were in the air. Then, they rendezvoused and made their way to Okinawa. En route to the target area, both the ceiling and visibility were excellent. Nevertheless, as the aircraft got closer to Okinawa, there was a layer of clouds from 1,500-3,000. There were some holes in the clouds, but it still would be difficult to complete their bombing mission.[7]

One of VT-10's TBMs in action over Okinawa. It is waiting for a strike assignment from the aerial coordinator. (USN photo via Arlo E. Schroeder and Steve Moore)

*****

I remember our Okinawa air strike on Easter Sunday. Easter came early that year. While flying, I thought that this was the day that Jesus rose up and we were going to go in and kill a bunch of people. That's the nature of war, but they picked the fight, we didn't. Our targets were some outlying islands of Okinawa, which were support bases. On our first strikes, we had to prevent the enemy from utilizing an airfield there. We bombed the revetments, which had some aircraft. Our fighters already did a sweep and there was no opposition in the air. So, we pot-holed the runway, blew up some hangars and storage facilities. We scattered bombs across the revetments. I had a dozen 100-pounder "daisy cutter" fragmentation bombs. There also was one 500-pounder general purpose bomb on each wing. These bombs spread scrap iron all over the place.

On the next hop, we attacked a mountain slope area which was emitting a lot of anti-aircraft fire. So, our fighters carried these five-inch HVAR's (High Velocity Aircraft Rockets). They were like five-inch cannon shells on the front of a rocket body. The Corsairs would carry eight of them. There also was another one called a "Tiny Tim," which was an eleven-inch cannon shell in front of a rocket body.[8] This was one that you had to point, but they were pretty good with them. I would see these guys flying wing on my aircraft and they would fire them right in front of me.

Yamato Attack

With the Allies on Japan's doorstep with their invasion of Okinawa, the Imperial Japanese Navy took drastic measures to fight back. Their response to this invasion was Operation Ten Ichi Go *(Heaven Number One). This plan called for the legendary battleship* Yamato *to sail from Japan with nine escorts, the light cruiser* Yahagi *and eight destroyers. The Imperial Japanese Navy considered this mission to be "one way" and it was essentially a sea-borne suicide operation. These ships were suppose to engage the enemy invasion forces then beach the* Yamato, *which would turn into a land-based fortress with its massive nine 18.1 inch main turret guns. The other purpose of sending* Yamato *and its escort ships was to divert the attention of allied aircraft so that land-based Japanese Army aircraft could do their* kamikaze *attacks against the invaders off Okinawa.*

Facing this small Japanese fleet was the large armada of the United States Navy Task Force 58, which was supported by a large compliment of ship-based naval aircraft and aviators. As Young mentions in his account, during the late evening of 6 April, Yamato *and it escorts were spotted by two American submarines, first the USS* Threadfin *and later by USS* Hackleback *but neither of them can get close enough to attack. Unfortunately for the Japanese, both submarines radioed detailed reports to COMSUBPAC at Guam and this information was passed on to the ships of Task Force 58 which were near Okinawa so they could prepare accordingly.[9] At 0832, the following morning, this group of ten Japanese ships was spotted again during a patrol by a Grumman F6F Hellcat from the USS* Essex. *Then, at 1014, they are sighted once again by two Martin PBM Mariner patrol bombers. As information trickled in about the details of these enemy ships, including the mighty* Yamato, *the die was cast and their fates would be sealed after being attacked by waves of ship-based United States Navy aircraft.*

*The American onslaught began between 0915 and 1015 with the launch of 280 aircraft from Task Groups 58.1 (*Belleau Wood, Bennington, Hancock, Hornet *and* San Jacinto*) and 58.3 (*Bataan, Bunker Hill, Cabot *and* Essex*).*

The legendary Imperial Japanese Navy battleship Yamato, *in 1942, while at anchor at Truk Atoll. (T. Kamino Collection)*

The strike aircraft for this first wave consisted of 132 fighters, 50 bombers and 98 torpedo planes. However, a group of fifty-three planes from the Hancock *launched fifteen minutes late then failed to find their target.*[10]

Next up were the 106 strike aircraft from Task Group 58.4 (Intrepid, Langley *and* Yorktown*) which start launching from their respective carriers at 1000.*[11] *Young and the rest of the Air Group Ten contingent (forty-two planes) from* Intrepid *were part of this second wave of aircraft which were going after the* Yamato *and its nine escorts. However, they did not begin to take off from their carrier until forty-five minutes later. The overall commander for the Air Group Ten aircraft that day was Lieutenant Commander Will Rawie (VBF-10) while the flight of twelve TBM Avengers from Torpedo Ten was led by Lieutenant Commander John Lawrence.*[12] *Young's crewmen for this historic mission were his usual crew (for this cruise), AOM2c John Floyd and ARM 2c Bill Kolodzik. These airmen flat-out refused to fly with any of the other pilots in their squadron. As Young recalled, "We had survived through a lot of dangerous flying and were a real team."*

*****

I HAD BEEN ASSIGNED to do the dawn anti-submarine patrol (that day). When I finished up my four hour flight and landed, the men were loading up the aircraft with torpedoes and bombs. They grabbed my aircraft, refueled it, and then loaded a torpedo aboard. There was a message to get to the Ready Room ASAP for a briefing. One of our submarines was in an estuary leading to Kure Bay and spotted a large group of Japanese ships, but he was smart enough not to attack it himself. As soon as they went south, he sent a message and gave the alarm. We were at "General Quarters" so during that time they were sending up the ordnance elevators from the bowels of the ship with bologna sandwiches, tomato juice, etc. This was our lunch, and at that time, the weather was pretty good. However, the weather forecast said that we would have squalls and we did. We had our briefing, then, launched, rendezvoused quickly and headed off. Lieutenant Commander Rawie in his beautiful Corsair raced ahead and spotted the enemy ships. (By that time), they were already hit by two other task forces like ours. I don't know how many of our aircraft attacked before our group, but I heard that the *Yamato* had already been hit by six torpedoes. Also, who knows how many armor-piercing bombs may have hit that ship; they shellacked the hell out of it.

We flew on, but the weather got worse and worse. We were in and out of the rain with a ceiling between 3,000-8,000 feet and low clouds. My skipper (Lieutenant Commander John Lawrence) plowed right through everything and never deviated from his course to reach the enemy fleet. So, we tucked in tight and flew on. I was lucky because my squadron was in the third group to hit this fleet.

*By the time Air Group Ten and the other strike aircraft of Task Group 58.4 arrived over the group of enemy ships (*Yamato *and its surviving escorts); they already suffered a lot of carnage as a result from the previous waves of air attacks. The Imperial Japanese Navy had lost the destroyers* Hamakaze *and* Asashimo, *which were sunk. The light cruiser* Yahagi *was dead in the water after being hit by a torpedo. As for* Yamato, *it was already hit by two 1,000-pounds and a torpedo from the first wave of American planes, which was followed by a second attack by a group of Helldivers which inflicted more damage on the battleship with their dive-bombing attacks.*

Each of the other two groups that hit (the enemy ships) claimed the final hit on *Yamato*. They had the final hit (maybe during their attacks) but it was still floating when we got there. Rawie (the overall commander for this strike) radioed and said, "That battleship is limping along and only doing about five knots. He is listing about five degrees to starboard and maintaining his course." It was good to have this knowledge ahead of time. We were in and out of squalls so I could see the enemy ships, then, I couldn't. So, we came in and our tactics were to split in half, then attack. Half of the dive-bombers went to one side (for their attacks); the other half of them attacked the other side. The same thing was done with the torpedo bombers. Six torpedo bombers attacked from the starboard side and I was (in the group) that hit the port side (left). I said, "Listen, let's leave the battleship for the submarines to get; they are up ahead and waiting for them. We should concentrate on getting those other ships, the cruisers and the destroyers." I had a torpedo aboard that was going to run at 22 feet, which wouldn't be too good against destroyers that only draw 18 feet. However, I never got into a position to drop on the destroyers because of the cloud cover and rain. I was on the port side of the force with half of my squadron's airplanes and on the far right of those six.

After Lieutenant Commander Rawie said go, we broke in. Well, here was this black smoke (anti-aircraft fire) in front of me so I pulled into the clouds in order to hide. So I went a little further and let me tell you I screwed up big time. I didn't come out of that cloud soon enough. When I did, I was right inside of the destroyer screen. I saw that the other aircraft in my squadron had already dropped (their torpedoes) and they were gone. One of torpedoes went ahead of the battleship and I think it hit a destroyer on the other side. I thought, here is the battleship and it is right in my face. So I didn't want to turn around and start again. I just punched over, stop-cocked the throttle and lowered my altitude to drop the torpedo. Previously, Rawie said that the ship was going about 5 knots and heeling to the starboard about 5 degrees (list). I set up all of this in my director and picked my angle. I kept jinxing my TBM until it came time to stabilize and drop that torpedo. You had to stabilize for ten seconds because the torpedo is steered with a gyro. 1, 2, 3, ---10, pickle! (Until you dropped the torpedo), you had to fly straight and level at 200 feet, then start escaping. Well, we were close, only about 1,000 yards, from the battleship. I figured that were no destroyers to the right and peeled off (in that direction). My turret gunner John Floyd hit the bridge area of that ship and we escaped. I was jinxing my TBM all over the sky again and stayed low (because it was harder for them to hit me). I raced out pass the stern of the ship because there were no destroyers behind it

Floyd who was sitting in the turret could see the bubbles from the torpedo at the beginning of its run. He said, "I don't know if we will hit it, No, I think we are going to hit it around the anchors." Later he said, "I think, we will hit it around mid-ships." After I was way out of there, he said, "There it went. We hit it." He could see the bomb blasts on the top-side of the ship. They got in just a little ahead of me. Our fighters had shot their eleven-inch rockets at them, which was a fearful weapon. But, it was a hell of an armored ship. Lieutenant Commander Rawie was still there and saw this lone torpedo bomber racing away. Later on he said, "What the hell did you think you were doing in there anyway?" Well, I thought that it was the only thing left to drop on. I didn't want to go and do a 360 for Christ sakes when everyone was already gone. Rawie witnessed that hit and took some photos with his hand-held camera. He said that when the explosion occurred, there was a big spray of water and smoke came up. Rawie thought that since the

Even though this picture shows an Avenger dropping a torpedo in training, it gives a good perspective of what Young's mission entailed during the attack on Yamato. *(NA)*

ship was cocked over at 5 degrees, my torpedo may have hit underneath the ship's armor and that it hit around the second turret.

(After this) there was nothing left to do so I prayed. I asked for help and got help. My radioman said that the right wing tank was leaking fuel so I shifted to that tank first to use up the gas. There was oil leaking from the engine but it ran smooth all the way back to the *Intrepid*. They held me off because they thought I would crash. So I was the last guy aboard. The gunner and radioman took out their gear and the aircraft was pushed over the back end.

This could be the view that Lieutenant (junior grade) Grant Young and his crew saw when Yamato *exploded. Going down with the mighty battleship were 3,056 men, which was around double of the people lost when RMS* Titanic *sank on 15 April 1912. (USN)*

*****

Remarkably, all twelve TBMs and the thirty-six airmen who flew them on this historic mission returned safely to Intrepid. *For their actions that day, four members of VT-10 were awarded the Navy Cross. The recipients of their prestigious award were Lieutenants (junior grade) Grant Young, Orbie Guthrie, Sid Overall and Lieutenant Commander John Lawrence. Out of these four airmen, Young was the only one credited with hitting* Yamato *while the three others were recognized for their successful air attacks on the battleship's escorting vessels.*

Yamato, *the pride of the Imperial Japanese Navy, was struck by at least five 1,000 lb. bombs and ten torpedoes.*[13] *At 1423, on 7 April 1945, after successive waves of American air attacks, the No. 1 magazine on the battleship exploded and a huge mushroom cloud rose up into the skies. This cloud, which could be seen 100 miles away, marked the end of* Yamato. *After this, it sank and there was a subsequent underwater explosion. The sinking of* Yamato *resulted in some horrific losses, estimated at 3,056 men, including the skipper Captain Kōsaku Aruga and Vice Admiral Seiichi Itō.*[14] *Furthermore, an additional 1,187 lives are lost during the sinking of five escort ships and the damage to the four remaining destroyers.*

The four VT-10 aviators who were awarded the Navy Cross, for their torpedo hits on the Yamato *and its escorts. They are standing in front of an Avenger, an aircraft which played such an important role in the Pacific War for naval aviation. Left to right: Orbie Guthrie, Grant Young, John Lawrence and Sid Overall. Lawrence was a lieutenant commander at that time while the other men were lieutenant (junior grade) in rank. Remarkably, Young was the only one out of these four men to hit* Yamato*, the other three aviators hit escort vessels during their attack. (USN via Grant Young Collection)*

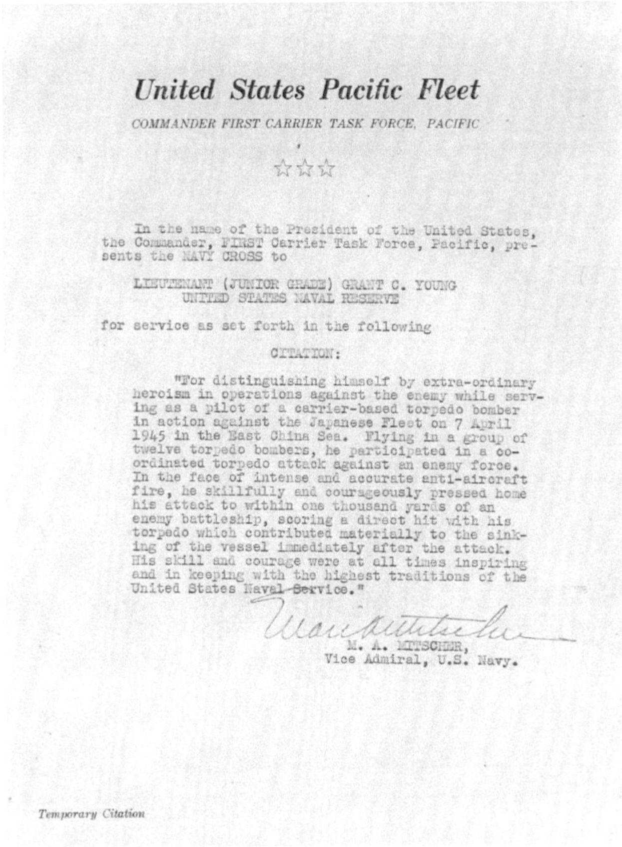

United States Pacific Fleet

COMMANDER FIRST CARRIER TASK FORCE, PACIFIC

☆☆☆

In the name of the President of the United States, the Commander, FIRST Carrier Task Force, Pacific, presents the NAVY CROSS to

LIEUTENANT (JUNIOR GRADE) GRANT C. YOUNG
UNITED STATES NAVAL RESERVE

for service as set forth in the following

CITATION:

"For distinguishing himself by extra-ordinary heroism in operations against the enemy while serving as a pilot of a carrier-based torpedo bomber in action against the Japanese Fleet on 7 April 1945 in the East China Sea. Flying in a group of twelve torpedo bombers, he participated in a co-ordinated torpedo attack against an enemy force. In the face of intense and accurate anti-aircraft fire, he skillfully and courageously pressed home his attack to within one thousand yards of an enemy battleship, scoring a direct hit with his torpedo which contributed materially to the sinking of the vessel immediately after the attack. His skill and courage were at all times inspiring and in keeping with the highest traditions of the United States Naval Service."

M. A. MITSCHER,
Vice Admiral, U.S. Navy.

Temporary Citation

Lieutenant (junior grade) Grant Young's official citation for being awarded the Navy Cross for his successful torpedo attack on 7 April 1945 against Yamato*. The Navy Cross is the second highest military decoration of the United States, with the Medal of Honor being the top honor bestowed by the country. (Grant Young Collection)*

The *Kamikaze* Strike Back

During the Battle of Okinawa, the kamikaze *(or suicide attack) air strikes were a constant threat and concern for any allied ships in the area. Unfortunately, on 16 April 1945,* Intrepid *would be struck by one of these lethal attackers during the battle.*

*****

Of course, we were worried about *kamikaze* attacks and they kept us "on edge" all of the time. We got up about a half an hour before dawn and the ship went to "General Quarters" because that was the time that they were likely to attack. The Japanese were not much at night flying, but they did have some night fighters at the end of the war. Finally, we heard that the one of our carriers, *Bunker Hill*, got hit.

During the *kamikaze* attack on our ship, it was hit three times before the fourth one hit our flight deck and blew us up. On other occasions, our gunners shot them down. The pilot who hit our flight deck must have been an accomplished airman because he did a split-S then hit the back end of our ship. Personally, I wondered what they were thinking about; dying like that. Instead, they could have attacked us, escaped then do it again. Our fighter pilots were really knocking them down, right and left. A lot of *kamikaze* pilots were just barely able to fly their aircraft. They would come in a "big herd" with escorting aircraft doing the navigation (to our position).

On that particular day, I had been up in the Ready Room so damn long, ever since dawn. They were coming in, wave after wave. I heard the 40s firing and the 20mms cutting in, now and then. So, I wandered out to one of the gun tubs, outside on the catwalk. I was talking to the men firing one of these 40mms. Every now and then, you would see a *kamikaze*. Aircraft would blow up or spin into the water. Suddenly, I heard "Holy Shit" and he was trying to get his guns pointed upward to shoot almost straight up. I looked up there and saw this little speck coming down on us. I panicked and wanted to get under the island were there was more iron to protect you. I started across the catwalk and only got about halfway across when the bomb blew up. I was amidships and there was a tremendous explosion. The blast blew me right against the overhead. The last thing that I remember was that it sounded like a

A perspective of the kamikaze *attack on* Intrepid *on 16 April 1945 as taken by a crew member from the USS* Alaska. *The ship in front of the damaged carrier is an American destroyer. (USN via NA)*

handful of marbles was thrown on the surface that I was walking on. I was (knocked) unconscious and when I came to, there was smoking coming down. Then, I came up to this metal partition; we called them "knee knockers." You have to get up over these partitions. When I got to this one, I couldn't feel my legs. So, I pulled myself up with my arms and discovered my legs were still there. I crawled on my hands and knees to the ladder so I could get some fresh air to breathe.

Meanwhile, the guns were still firing. But, the fires were still raging underneath us. This ammo had to be thrown over the side. I don't know how I managed to handle the five-inch shells the way I did; they were heavy. We emptied the magazine and the fire was raging right underneath us. We had to empty the magazine so it wouldn't blow up. This Boatswain's Mate says, "You're bleeding. You better go see the medic. This guy took the forceps then yanked out the shrapnel fragment sticking out from my scalp. I felt like I was drunk then started to sober up. I still don't know how I managed to throw those cannon shells, but I did. Later, I received the Purple Heart.

Some of the carnage on the Intrepid's *hangar deck with a destroyed Corsair fighter in the foreground. (NA via Steve Moore)*

*****

As a result of this suicide air attack against Intrepid, *seven men perished outright, one was missing and another man died later succumbed for his wounds. Therefore, nine men were killed. Wounded in Action consisted of one officer and seventy-one men.*[15] *The Japanese aircraft hit the flight deck about thirty feet from the Ready Room. The bomb on the plane blew up the hangar deck (below the flight deck). The crew on board the carrier worked quickly to push aircraft overboard and get the fires under control. Remarkably, the men were able to install a temporary patch on the flight deck, which enabled the CAP flying overhead to land back on the ship.*

Unfortunately, it was necessary to properly complete the repairs on the damaged Intrepid, *which went back to Ulithi on 18 April. The carrier arrived there the following day but on 4 May, it was ordered back to Pearl Harbor for additional work. Next, the* Intrepid *sailed back to the mainland and was moored at Naval Air Station Alameda. From there it went to Hunters Point and entered the dry dock on the 27th so that repairs could be completed. As for Young and his buddies, they were happy to get a stateside two-week leave.*[16]

By 4 June, all of the men had returned to Naval Air Station Alameda and they were ready to get back to work. For the next three weeks, the airmen had to test fly their brand-new TBM Avengers and do some training hops with them. On 29 July 1945 at 1330, the repaired Intrepid *complete with her crew left port bound for Pearl Harbor. It arrived in Hawaii on 5 July and two days later,* Intrepid *did a four day shakedown cruise around the Hawaiian Islands. Next, Young and the rest of Air Group 10 went to Naval Air Station Barbers Point and did some training missions. They conducted both torpedo and rocket practice at this time.*

On 30 July, Intrepid *finally left Hawaii and was bound for Eniwetok Atoll in the Marshall Islands. The revitalized carrier was escorted by two destroyers for the journey. During the trip to Eniwetok, on 6 August, Air Group 10 would fly its last combat missions of the war against the Japanese military installations on Wake and Peale Islands. The group did four fighter sweeps and two bombing strikes that day. Young flew on both bombing missions that VT-10 did. In summary, these combat sorties resulted in the dropping of forty-two tons of bombs and the firing of 490 five-inch rockets against these respective targets. With these missions completed,* Intrepid *continued on to Eniwetok Atoll and were anchored there when hostilities against Japan officially ended.*[17]

Lieutenant (junior grade) Young (left) and his buddy Walt at a discharge celebration held on 23 June 1945 (not for Young though). (Grant Young Collection)

Some Reflections on Flying

During my conversations with Grant Young, I spoke with him a great deal about his "flying" experiences and how he felt about some of the respective aircraft he flew.

*****

Regarding skills to be a good pilot, you had to have excellent "hand to eye" coordination. You had to have the ability for long-range vision and calm nerves. It was important to have the ability to think under pressure and make instant decisions. You always had to keep your eyes open and swivel your head (look behind your aircraft). Maybe I was just lucky to be raised on a farm in Lee County, Illinois. I was used to having responsibilities and working with my hands. Because I worked with machinery, I was careful where to put my hands and feet. My father taught me to be cautious around whirling shafts and flying things. I was mechanically-inclined and maybe that is part of being a pilot. After a certain amount of experience, you could sense what was happening with the machine.

All during World War II, I flew the Grumman TBF/TBM Avenger. I was one of the few people that crossed over from torpedo-aircraft then to dive-bombers, next to prop-fighters and lastly, jet fighters. I was just lucky that I got to make carrier landings in all of them. I thought the Avenger was a magnificent aircraft and flew two combat tours with it. Sometimes, I was hit by shrapnel several times, but my aircraft was never damaged so badly that it could not bring us home. The plane had heavy armor plate around the pilot seat and self-sealing fuel tanks. For example, during my attack on the battleship *Yamato*, we were hit and the fuel tank in the right wing started to leak and did so until we landed. It was a stable aircraft to fly, but very heavy so you didn't want to do aerobatics with the Avenger. However, I did a big barrel roll with it, just to see if I could! It really wasn't designed to do aerobatics or vertical dives, like dive-bombers do. The plane was beautiful and smooth to fly. It was a heavy aircraft and kind of heavy on the controls. With plenty of altitude, I did barrel rows in the Avenger. I tried slow rolls with it, but the aircraft always pointed down. It practically landed itself. I used to crash down with them (on the deck) and never had any trouble with the gear folding in on them.

Four VT-10 TBMs on their way back from a mission, unfortunately, there were no other details of this picture. (USN via Grant Young Collection)

Early on our torpedoes were poor, but the aviation ones were not as bad. It was the submarine guys who really had the troubles with them. Personally, I carried torpedoes twice. The first time was when we went to Yatsushirō Bay (on 29 March 1945), but there were not many ships there. We just found one lonely transport so the skipper ordered a couple of guys to attack it. However, the torpedoes weighed 2,800 pounds each and (the rest of us who didn't use them) could not return to our ships as is. So, there was this big bridge over an estuary there and I targeted it along with a few other guys. I do not know who hit it, but the bridge blew up.

200 feet was the desired altitude to drop a torpedo from. During this time, you had to use a lot of power and ideally be flying at 250 knots. Torpedoes have gyros mounted in them. Typically, you are jinxing your aircraft all over the sky to avoid anti-aircraft fire but there comes a time when you had to fly straight and level to drop your torpedo. You had a torpedo director that sat on top of the dashboard and you had to calculate the speed of the vessel that you were dropping on along with your angle for attack. Then, you had to swing that around to determine what direction you had to be flying to get a lead on the ship. You had to do that and stabilize for about ten seconds so the gyro and torpedo could remain steady. Therefore, if you were skidding, the torpedo would do the same thing. So, it was that one spot when you were really in peril and you had to settle down for ten seconds. After you dropped the torpedo, your aircraft was as light as a feather then the doors slammed shut.

(In my postwar years), I did thirty carrier landings in the Grumman F8F Bearcat. For prop-driven fighters, the Bearcat was the best. It also was the best aircraft that I ever flew. It felt like I was driving a fancy Corvette. But, the R-2800 engines were gas hogs and the fighter didn't have any range. The F8F was built to jump off and shoot down the *kamikaze* aircraft.

Inter-War Period

Between the wars (1946-1950), Grant flew as a reserve ferry pilot then joined Attack Squadron 95 (VA-95) and Fighter Squadrons 92 and 74 (VF-92/VF-74). While serving with these units, the young airman flew from the following aircraft carriers: USS Philippine Sea *(CV-47), USS* Leyte *(CV-32) and USS* Kearsarge *(CV-33).*

Post-World War II picture of the now-experienced naval aviator, Grant Young, and a Grumman F9F Panther jet. He flew the photo-reconnaissance version of this aircraft during the Korean War. (Grant Young Collection)

Korean War

I GOT INTO JET fighters at the onset of the Korean War and was posted to Reconnaissance Squadron 61 (VC-61) on board the USS *Antietam* (CV-36). I flew with Howard Crews, my first "hero" in naval aviation. I was lucky enough to be assigned to his squadron. Crews was my commanding officer and I was a lieutenant at that time. My squadron had Grumman F9F Panthers and later on when I was finishing up my tour, the McDonnell twin-engine Banshee. The Panther had a Cadillac airframe; very maneuverable, but it didn't have enough power. If you got above 25,000 feet and went into a heavy turn, the fighter didn't have enough power to push you around. My carrier was made for non-jet aircraft. So, when we put jets on those World War II era carriers, we were putting maximum stress on all that equipment. The 95-foot hydraulic catapults launched us to 125 knots. This was a thrilling experience but it was nice using steam catapults later on. It was dangerous flying those F9Fs off that type of carrier. However, that is what we did. Sometimes the "old boat" could churn out about 27 knots. They claim in order to do this; the chief engineer had everything open but his fly! Therefore, they would take a couple of rockets or 250-pound bombs off (to lighten up the aircraft). Then, you would take off and it still would be buzzing, which would be a stall warning.

On several occasions, I had to go to the Yalu River towards Manchuria to take photos. They took out the bottom two 20mm cannons in the nose to mount the camera, which was vertically mounted. The trouble was that the magazine was so snug against the barrels of the 20mm cannons above. The heat would scramble the film. So, they told us to not to fire the cannons unless we were in an emergency. But, we always had a fighter escort (to protect us) when we up towards the Yalu River. They would give us a division escort of the "regular" (unmodified) Panthers. It was a great aircraft; stable to fly, but it did not have an auto-pilot. We didn't have a view-finder in the aircraft so you had to judge where you wanted to point your camera. The Panther had an oblique-mounted camera that would fire out in a 45-degree angle. At that time, it was our primary jet aircraft for photography. Previously, they had been using the Corsair and the Hellcat. However, we never sent the Hellcats out on missions; they were used only for training. Corsairs flew on missions when the Korean War started. I was in the third detachment and flew only jet aircraft.

At 25,000 feet, I could out-turn the MIGs. They had evidentially had encountered us before and would be up there at 35,000-40,000 feet. However, I would be flying at 25,000 feet in order to do aerial photography (to get good details). So, a couple of times I saw them up there but by that time they must have been low on fuel and to return home. I was lucky that I never had to do more than turn with them. They left me alone and all of my comrades too.

The Panther was a stable, great aircraft. With four 20mm cannons you could tear up a lot of things. Once, we came to this dam and roaring across it was a motorcycle and side-car. The motorcycle had two people in it, but we were not sure what side they were on. Another time, I saw two ox-carts. Typically, they were used to carry ammunition or food. The two ox-carts were parked by a shed. I saw this figure run towards the ox-carts just as one blew up. I saw some of this junk (debris) through the air just as it blew up.

Test pilot days

I WENT FROM VC-61 to become a test pilot at the Naval Air Test Center NAS Patuxent River, Maryland in October 1952. There I got to fly everything in the barn from the big P2Vs to whatever came down the pike, which was being researched. In those days they said, "Here is the handbook on it, go fly it!" But, in a larger aircraft, I went with someone who was more experienced (flying) it. If you have a place for two pilots, it is a good idea to have someone along who knew what they were doing. I also wanted a good checkout.

Pentagon Duty & Return to Flying

In April 1954, Grant Young transferred to the Offices of the Chief of Naval Operations (in the Pentagon, Washington, DC.) and served there until May 1956. Then, that same month, he became Assistant Navigator of the USS Antietam *(CVS-36). Young became so proficient in this job that he felt like it was a 14' motor boat. Many times he was just one of the three qualified officers of the deck when it was underway.*

Thankfully for the veteran airman who enjoyed flying so much, in January 1958, he was reassigned to active flying duty in Attack Squadron 172 (VA-172). During the two years Young was assigned to this squadron, he did two cruises in the Mediterranean Sea on board the USS Franklin D. Roosevelt *(CVA-42). Next, in January 1960, he transferred to COMFAIRJAX, as an Admiral's staff member who had the responsibility of supporting and monitoring the training activities of every naval aviation unit based in the Jacksonville, Florida region.*

In September 1961 Grant was given a sabbatical leave for educational purposes. He was sent to The George Washington University (Washington, DC.). There he completed his BA degree, which was designed by the Navy to equalize career and promotion opportunities for all officers.

Cold War Tensions

My last flying assignment was with Attack Squadron 34 (VA-34). In February 1963, I was posted to this squadron as commanding officer. Our basic "mission" was to deliver nuclear weapons so (prior to this) I had to attend a three month nuclear weapons school. I did two cruises again in the Mediterranean Sea, on the USS *Saratoga* (CVA-60). I was qualified to monitor the health of a nuclear weapon. I was the commanding officer of VA-34. During this time, Russia began to shadow us. The ship looked like a trawler, but it was a destroyer. It had a massive boom, that looked like it could carry a net, but this was a massive antenna. They were listening and recording all of the radio talk. Later on, they got to the point of trying to embarrass our skipper. Once you hoisted the "Fox flag," you had the right of way over everything except row-boats. This flag had a red square set 90-degrees in a white square. It meant flight operations were going on. They would try to embarrass us by cutting in front of our carrier. The captain would say, "I'm going to cut them in two." But, they were real ship-handlers those people. They would miss us by a couple of hundred yards, but we were required by law to maintain our course and speed. If you were flying the Fox flag everyone was suppose to recognize that. So, it was a tense time and there was always that threat of something appearing over the horizon and heading for Washington D.C., New York, etc.

Sometimes it was so tense that the admiral said, "We have to demonstrate to those people that I don't care what it is doing outside, we are ready to deliver a weapon or more, only in response." He would say pick a couple of A-4 drivers (Douglas A-4 Skyhawk) and send them along even at 2AM in the morning, while it was raining. I was one of those "hot shots." I had more instrument time than anyone else. I remember the times sitting in my little A-4 taxiing up to the catapult and I could not see because rain was pouring down on my windshield. Take off would be done with a 330-foot long steam catapult which would give you an even start off the ship. I would go off and fly three hours with my wingman then land in the same kind of slop. I did that a number of times just to prove we could and would.

Unfortunately, at age forty-five, I was too old to fly anymore. Therefore, in June 1965, I was ordered next to the U.S. Naval War College, Newport, Rhode Island, for post-graduate study in matters related to economic and political warfare.

Vietnam

In October 1966, I transferred to the USS *Kitty Hawk* (CV-63). I served the ship and its embarked air group during two cruises to the Tonkin Gulf/Yellow Sea area of the Vietnam conflict in their support of COM 7th Fleet activities and responsibilities. I was the Air Department Officer on the *Kitty Hawk*. I ran the flight deck, hangar decks and fuel farm. Anything above the hangar deck belonged to me. I had 750 people working for me. I felt the negroes on board were capable people and Big Boatswain Holland kept the peace. His favorite comment was, "If you don't shape up, you are going to fall down a ladder." On board the ship, we had ladders going deck to deck. We had a wonderful cruise. I never had any "race trouble" because I know that he threatened everyone "equally" and didn't discriminate. Our department ran so smoothly.

Last Navy Assignment

In May 1968, I was assigned (again) to the Chief of Naval Operations, Strategic Plans and Policy Division at the Pentagon. This was a very tough assignment. Frequently, I had to work fourteen hour days, often seven days a week. My office was responsible for, among other things, the contingency plans. They were capable of knowing twenty minutes before impact that a weapon was coming to hit the United States and we would know were it came from. Then, our B-52s with ours (nuclear weapons) would be on the way and our submarines would go to launch depth. Armageddon would happen. I was going to be a captain of a ship but they said I would have to stay there another year. I did not want to do this after already being there for twenty-six months. So, I retired at the rank of captain on 1 July 1970 after nearly thirty years of service.

*****

Grant C. Young dedicated a large part of his life to serving in the United States Navy and doing what he loved best – flying. Remarkably, he participated in World War II along with the Korean and Vietnam conflicts. In his peacetime service in the Navy, Grant did everything from test-flying to working at the Pentagon. In recognition of his naval service, he was awarded the following decorations: the Navy Cross, Air Medal with six gold stars, the Navy Distinguished Service Medal and the Purple Heart. The Navy Cross is considered the highest honor in the Navy. The squadrons Captain Young served in were awarded two Navy Commendation Medals, two Presidential Unit Commendations along with several War Campaign and Combat Action Theater medals.

Captain Grant Young towards the end of his long and varied career in the United States Navy. The young boy from a farm in Lanark, Illinois came a long way and surpassed his boyhood dreams of becoming an accomplished Navy pilot. (Grant Young Collection)

One of the several visits with Grant Young was with my good friend Mark Stone (left), a passionate naval aviation history buff. We felt so fortunate to have met and befriended this wonderful Illinois veteran. (Ron Werneth)

Family/Retirement

We own over 160 acres in Lanark, Illinois but don't farm it anymore. I retired from farming when I reached Medicare age (sixty-five years old). In the winter of 1986, I promised my wife that I would quit farming. I had a big farm sale and sold all my beautiful green machines at auction. In 1987, I rented my land out to a young man and he has been farming it ever since.

WW II Feelings

I regretted very much that we had to fight Japan but I was willing to. I was a pretty "raw cut" farm boy and didn't have any college. I enlisted and passed the exams to go to OCS (three-month program at the University of North Carolina). This was a wonderful adventure. Right from there, I went to flight school. I had a background in Internal Affairs and Relations. But, I wasn't aware of the failures of the diplomats on both sides, prior to the war. They (Japan) picked a fight with us at Pearl Harbor. We were going to kill them until they would sit down with us and talk peace. That is what war is about. Sometimes, that is what our politicians and diplomats forget. If you try to short-change this process, you will have a stalemate, like we had in Korea. The same thing happened in Vietnam. This is because they would not let us prosecute a war like it should be fought. They wanted to do it "humanely" - there is no humane way.

Grant Young passed away on 16 August 2015, preceded in death by his sweetheart Ethel.

[1] Young's aircraft, a TBF Avenger torpedo bomber, Bureau Number 06064, was located on 5 May 1993 by A&T Recovery. The Chicago firm that found it specializes in the recovery of WW II aircraft. Unfortunately, Grant Young's TBF has not been recovered yet from Lake Michigan. http://atrecovery.com/Pages/CaptYoung.htm

[2] The Super Chief was the famous passenger train that was operated by the Santa Fe Railway and ran between Chicago, Illinois and Los Angeles, California.

[3] The background information for this section was gathered from primarily one resource: Stephen L. Moore with William J. Shinneman and Robert Gruebel, *The Buzzard Brigade: Torpedo Squadron Ten at War*, 140-156.

[4] While still flying with VT-51, on 2 September 1944, Bush piloted his TBM on a strike against an enemy radio station on Chi Chi Jima (Father Island). His crew consisted of Radioman Second Class John Delaney and rear gunner Lieutenant (junior grade) William White, USNR. During his bomb run, the engine on their Avenger caught on fire due to intense anti-aircraft fire from the ground. However, Bush still dropped his bombs before flying several miles away from the target area (above the water). Sadly, only Lieutenant (junior grade) Bush survived the bailout. Later, he was rescued by the submarine USS *Finback* (SS-230) and spent one month on board before eventually making his way to his carrier, the USS *San Jacinto*. Later, the future 41st U.S. President was awarded the Distinguished Flying Cross for his action and heroism on this combat mission.

[5] Moore, Shinneman and Gruebel, *The Buzzard Brigade: Torpedo Squadron Ten at War*, 225.

[6] Some reflections of John Hyland, who later became an admiral from Admiral Ronald J. Hays (USN. ret.). "I knew Admiral John Hyland well since I was his strike warfare officer for about a year when he was 7th Fleet Commander. We were aboard his flagship, the cruiser USS *Providence* (CLG-6), and spent most of the time in the Gulf of Tonkin fighting the North Vietnamese. He had a great sense of humor, approachable, well liked by his crew, smart and a respected leader. After retirement he lived in Honolulu so that he could play tennis every day."

[7] *The History of Torpedo Squadron 10: 15 September 1944 - 26 November 1945 (Unofficial)*, 20-22.

[8] This lethal ordnance was an 11.75" air-to-ground rocket that had a 500-pound semi-armor-piercing bomb with a maximum range of 1,500 meters (1,640 yards).

[9] ComSubPac stands for Commander Submarine Force, Pacific (USN).

[10] Rear Admiral Samuel Eliot Morrison, *History of the United States Naval Operations in World War II, Vol. 14: Victory in the Pacific*, 197.

[11] Vice Admiral Marc A. Mitscher, *Task Force 58, Action Report for 7 April 1945*, 9. Please note other sources list different numbers of aircraft launched during this second wave, however, Vice Admiral Mitscher's report should have the accurate figures. Furthermore, the author has seen other published works that state the carrier USS *Independence* (CVL-22) contributed aircraft for this strike, but this is not the case because it was separated for refueling at that time.

[12] *Air Group Ten Action Report: 7 April 1945.*

[13] www.combinedfleet.com

[14] While researching *Yamato* and the sinking of the ship, I discovered different numbers of men who perished. My sources included the original Japanese war records and numerous other official sources. However, I believe the number of 3,056 men lost should be fairly correct as this is the figure given by the *Yamato* Museum after careful research by their staff.

[15] *The United States Navy Casualty Listing for World War II* (kindly provided by Jim Sawruk).

[16] The background information on the aftermath of *Intrepid's kamikaze* attack was derived from two sources. *The History of Torpedo Squadron 10: 15 September 1944 - 26 November 1945 (Unofficial)*, 28-29 and Moore, Shinneman and Gruebel, *The Buzzard Brigade: Torpedo Squadron Ten at War*, 281-285.

[17] Ibid. 285-288. The background of these two paragraphs was derived from this invaluable resource.

Chapter 10

"As a soldier, you did what you were told to do; it was just a job."

First Lieutenant David R. Braden (USAAF)

Since my days as a high school student, I was always fascinated with the Boeing B-29 Superfortress and its significance in ending the war over Japan in the Pacific. This long-range heavy bomber was considered state-of-the-war over seventy years ago with pressurized crew compartments and remote control guns. It is well-known as being the one that dropped the atomic bombs on the Japanese mainland, which became a controversial topic many years after the war was over.

During one of the symposiums that I attended at the National Museum of the Pacific War (Fredericksburg, Texas), I was able to meet local Dallas-based former B-29 navigator David Braden, who at that time was serving as the president of the Twentieth Air Force Association. I met David and his wife Sara; both were very kind, articulate and classy. During the Pacific War, this distinguished aviator completed the required combat tour of thirty-five missions which included dangerous low-level fire bombing raids, surviving a crash-landing and several other perilous experiences.

Early Years

I was born on 10 November 1924 in Dallas, Texas. When I was twelve years old I visited Randolph Field in San Antonio with my uncle. From that moment on, I knew I wanted to fly and that the Army Air Corps was the place to learn. I was a seventeen years old aeronautical engineering major, college freshman, when Pearl Harbor was attacked. As soon as I was eighteen, I enlisted in the Air Corps College Reserve Program. Shortly, thereafter, I was called to active duty; it was all a series of logical choices for me.

After being called up in November 1943, I went to Sheppard Field, Wichita Falls, Texas, for nine weeks of basic training. Next, I went to the University of Tennessee for sixty days where I had some academic courses and ten hours of instruction flying a Piper Cub. From there I went to Nashville, Tennessee, and was classified as a navigator because I was near-sighted in one eye and far-sighted in the other one. So, I was very disappointed. Then, I had to wait a bit before being sent to Selman Field (in Monroe, Louisiana) to attend pre-flight navigation school. After this, they sent me to Florida for gunnery school then, I came to Monroe for advanced navigation school where I graduated as a second lieutenant.

Upon graduation, I went to Boca Raton, Florida, for a class to study using radar for navigation and bombing. Next, I went to Clovis, New Mexico where I was assigned to a combat crew. I was a member of the 870th

Bomb Squadron, 497th Bomb Group, 73rd Bomb Wing. From Clovis I went to Nebraska and our crew went to Saipan via Hawaii, then Kwajalein Atoll before reaching our destination.

B-29 Reflections

At that time, the B-29 was top secret. When I was in navigation and radar-bombing schools, I did not know why I was being taught radar-bombing. While there I found out that I would be in a B-29, a top secret weapon, which I never even saw one before this time. When I got to Clovis, we still had to do some training in the B-17s because the B-29 was so new. We were considered to be elite crews. All of the aircraft commanders were either former instructors or veteran aviators who had already flown combat missions over Europe. As for the rest of us, many people were at the top of their graduating classes. One of the things that was interesting is when we flew B-29s on training missions, if we went anywhere, we had to post an armed guard with the aircraft; it was a real honor. Of course, we knew that we would be going to Japan because that was the only plane that could fly there.

The B-29 was a superb aircraft. However, initially, these bombers were plagued with terrible engine problems and fires starting. They had a lot of that when the 58th Bomb Wing was in India. However, the Boeing Company finally figured out what to do by adjusting the cowl flaps on the engines. So, they had their engineers go to India to rectify this issue. Nevertheless, even when I flew in combat, we still had a lot of issues. The engines were not very dependable. In fact, on 3-4 missions, we had to abort because of losing an engine on the way up. Engine failures made take offs on a fully loaded bomb run very dangerous. But, it was suitable for combat as no other plane could fly the 3000-mile round trip; much less carry the bomb load we carried. The gun system was terrific as well.

Eighteen year old, aviation student David Braden at the Nashville, TN. In November 1943, he signed up for the Army Air Force Reserve aviation cadet program. Two months later, Braden was called up and sent to Sheppard Field for basic training. (David Braden Collection)

Braden at home, on his first leave, in Dallas, TX., as a newly promoted second lieutenant. (David Braden Collection)

First Lieutenant David Braden in his typical 20th Air Force uniform while serving on Saipan. (David Braden Collection)

Braden's first B-29 crew picture, taken after their arrival in February 1945 on Saipan. He is pictured in the center of the front row. Front row, left: Aircraft Commander, First Lieutenant Norman V. Westervelt and next to him is the bombardier, Second Lieutenant Gordon F. Nedderson. On 3 March 1945, both men were killed during their 3rd mission together after they ditched their damaged bomber. Back row, left: CFC gunner, Technical Sergeant Robert L. Curtis, broke his back during this ditching, and was immediately sent back to the United States for recovery. Back row, third from the left: tail gunner, Sgt. Herman C. Knight, Jr. Far right, back row: radio operator, Staff Sergeant Gerald F. Frorillo. After being assigned to another crew, on 26 May 1945, both Knight and Frorillo were involved in another B-29 ditching (tail number A Square 38). Knight was rescued by the United States Navy submarine USS Pipefish *(SS-388) but Frorillo drowned. (David Braden Collection)*

Combat

On 28 January 1945, Braden arrived on Saipan and was based at Isley Field as a member of the 870th Bomb Squadron, 497th Bomb Group, 73rd Bomb Wing. The young navigator was part of an eleven-man crew led by aircraft commander First Lieutenant Norman V. Westervelt. While this crew was together, they flew on a Superfortress (serial number 42-63412), tail number A Square 25, adorned with the nose art of "Peace on Earth." After arriving, they participated in two missions during the month of February, and safely returned home. However, on 4 March, the crews' luck changed. The primary target for this ill-fated third mission was the Nakajima aircraft plant (in Musashino, part of Tokyo).

*****

We flew a high-altitude mission to Tokyo and were hit by flak in one of our wing tanks. When we turned off the target and started to head back over the ocean, the radar man went back to close the camera hatch at the rear of the aircraft. There, he smelled gasoline and saw that it was streaming all along the bottom of the plane so we knew right then, 1,500 miles away from home, that probably, we would have to ditch. Immediately, we started to throw anything that had any weight into the ocean including the Norden bombsight.[1] We got within 220 miles of Saipan and were constantly reporting our position every thirty minutes on the radio back to base because of the sixty percent of the B-29s that ditched in the war, the crewmen were never found. We didn't know our exact position but it was near Sarigan Island. The aircraft commander (First Lieutenant Westervelt) said that we would have to go down and the sea was very rough with waves that were fifteen-twenty feet high. The procedure for ditching at sea was to drag the tail in the water to slow the aircraft down and hold the nose up high. Therefore, as the waves approached, you would lift your nose over them until the aircraft slowed down so it could land gracefully. Well, just as we were going to touch the water, all four engines went dead. Of course, the plane just flopped down and we ran head on into about a fifteen foot wave, going around 100 miles an hour. This just crushed the glass nose of the aircraft, which immediately killed First Lieutenant Westervelt, who was decapitated, and bombardier, Second Lieutenant Gordon F. Nedderson. The central fire control man, Technical Sergeant Robert L. Curtis, had his back broken in the impact, but he and the rest of us (nine men, out of a crew of eleven) got out safely.

We went into two rafts, which were tied together for awhile. It was starting to get dark and we started to get worried because during the night, we would drift and get lost. Because of this, twenty-hour hours later, they would have to do a square search to find us. Just about dusk we heard a flying aircraft above; it was a Navy PBY Catalina flying boat. We called them "Dumbos," and they were used for doing air rescues. Quickly, we spread some sea dye in the water; he saw us, waggled his wings and came down. This particular crew had flown eighteen months and never had found anyone before us. They were getting ready to rotate home and risked their own lives while doing this. For me, the take off in the PBY was worse than ditching because when the aircraft would hit a top of a wave then bounce all over, come back down and hit another one. But, he finally got it off the ocean and when we got home there were ambulances waiting for us. They took everyone to the hospital for checkups, then, we all got shots of whiskey and ate steak and eggs. We also got beds with sheets in them.

The next day everyone went back to the squadron except Technical Sergeant Curtis, who was immediately sent home. We never saw or heard from him until many years later when a newspaper reporter found this veteran for us. After returning back to duty, we learned there was another crew in our Quonset hut that had ditched the night before and they had lost their navigator, flight engineer and radar man. So, three of our original crew took over these positions and joined this "new" crew. The rest of the men from my original aircraft transferred to other crews but everyone stayed in the same squadron.

*****

After surviving the 4 March crash landing essentially unscathed, Second Lieutenant Braden became a member of the eleven-man crew led by Captain James Buckheit. Just prior to this, Buckheit's first B-29 "Miss Margaret" had to ditch and his original navigator was seriously wounded during the crash-landing. Furthermore, as a result of this ditching, he lost his flight engineer, radar operator and one other crew-member. Braden and three other men joined the newly reformed Buckheit-led crew and their replacement B-29 (serial number 42-63466) was named, "Striped for Action," with A Square 32 on the tail.

On the night of 9-10 March, five days after Braden's ditching in the ocean, the first fire raid on Tokyo was done. It killed an estimated 83,793 residents. The fires started as a result of this attack burned 16.8 miles of the city down and left 1,008,000 residents homeless.[2] According to this veteran's recollections, he said that everyone thought that it would be a suicide mission especially since they were flying so low (5,000-7,800 feet). Since Braden and the others recently ditched, they didn't have to go. They ended up flying their first mission, as a new crew, a few days later.

Left: Nose art of B-29, A Square 25, "Peace on Earth" (42-63412), the first aircraft Braden flew on, which on 4 March 1945, was ditched at sea due to combat damage. (David Braden Collection)

Below: An amazing picture of "Peace on Earth" after ditching as seen from the air. (David Braden Collection)

Left: The Superfortress Braden flew the duration of his combat missions on after ditching on 4 March 1945, A Square 32, "Stripped for Action" (42-63466). The veteran's comment on this nose art, 'You can tell what we were thinking about a lot on Saipan!" (David Braden Collection)

Daily Life on Saipan

Our aircraft was "our" aircraft and nobody else flew it. Our ground crew had their names painted on the planes just like we did. They were "Mother Hens" to us. We were just a bunch of kids, but some of them were older than us. There was a strong relationship between us and they were waiting for our B-29 every time we came back. But, we never had anything like they did in England. There were really no social activities that we did together.

We did have an Officers' Club, but there was no refrigeration or ice. So, if you wanted to drink a warm beer or whiskey this would be your opportunity. We also had a movie theater called "Goat Gulch." It was on a hillside and barrels, which were turned on their sides, were our seats. There was a giant movie screen and every night a different movie would play, and everybody would go to see it. I would take my raincoat and helmet-liner because just about every night around twilight it would rain just a little bit and get me wet. So, when the rain went away, I folded up my raincoat and it became a cushion. Lastly, getting letters from home was a very big deal. I was engaged to my girl back home and later we got married. So, I thought about her a lot.

497th Bomb Group B-29s parked in their hardstands on Isley Field, Saipan (NA)

Some of the American facilities at Isley field, Saipan while Braden was stationed there. (NA)

Typical Mission Gear

I WORE THE FOLLOWING layers for the trip there and back: T- shirt, shorts, flight suit (coveralls) no electrical heat connection, socks (two-layers), shoes and fleece-lined flying boots. I also had a survival vest containing fish hooks and line, map of Japan and Pacific printed on silk, a compass, bird shot ammo for my .45 caliber automatic pistol, and a lot of other stuff I can't remember. I had a ball cap, sunglasses, throat mike and earphones for communication, an inflatable Mae West life preserver and a .45 pistol covered with a condom in a leather holster strapped to my leg. Furthermore, my gear included a .25 caliber automatic pistol, ammo, a 7" blade hunting knife in leather, shoulder holster, along with a parachute harness over all that was attached to a life raft as a seat pack

Over Japan depressurized, I added the following to the above: a flight jacket, a leather flight helmet with oxygen mask, goggles (interchangeable lenses, clear or green for daylight and red ones for night flights); and silk gloves, thin leather gloves along with heavy gloves that were worn in layers. I also had a heavy steel flak jacket which covered the entire upper body and a steel combat helmet with ear flaps covering the earphones.

I carried no I.D. except my dog tags, my father's dog tag from World War I, and a four-leaf clover, back to back with a picture of my fiancée, encased in plastic (about the size of a silver dollar) all on a chain around my neck. My parachute was a clip-on chest pack which rested on the floor next to my desk.

A group of 497th Bomb Group B-29s flying in formation on their way to attack mainland Japan. (NA)

An undated picture of 73rd Bomb Wing B-29s flying over Mt. Fuji, during a mission. (NA)

Missions over Japan

SECOND LIEUTENANT BRADEN'S initial combat missions were flown during the daytime but this changed to a different nighttime strategy later on.

*****

FOR DAYTIME MISSIONS, we had an announcement at the squadron level as to which crews would fly and aircraft number. We were up early for a group briefing for daytime missions which were flown at high altitude. We flew the same plane every time unless it was out for maintenance. Take off was one of the more dangerous aspect of every mission. Loaded with 7000 gallons of fuel, 10 tons of bombs, 7000 rounds of .50 caliber ammunition, we would roll down the runway through what the pilots called the "Valley of Death." This is the point where, if you lost an engine, you were either going too fast to stop or too slow to fly. In either event you fell off a 450-foot cliff at the end of the runway, into the ocean and certain death. Standard procedure was to wait until the last moment rolling down the runway until you could reach the speed to lift and fly, then immediately retract the landing gear, fly straight off the end of the runway and dive off the cliff edge toward the water to gain flying speed and more lift. We would pull up so close to the water we blew spray from the ocean on the plane. Your stomach would react like a drop in a fast elevator. It was very scary sitting in the "greenhouse" nose, but exhilarating in a crazy sort of way, especially if you like roller coasters.

Flying time took about seven hours to arrive in the afternoon. At first loose formation flying a 1,500 feet, then about 100 miles out from Japan, we rose to altitude (25,000 to 30,000 feet), tightening formation to fly to initial point then turn to target. There was flak and fighters all the way in. Peel off over target then split up after we left the coast and the fighters behind, then flying home singly, every plane on its own, usually arriving after dark, celestial navigation required, until later when we got Loran (a navigator's idea of a gift from God!)[3]

As an anecdote, flying at high altitude, Japanese anti-aircraft was effective on the bomb runs, but not nearly as effective or intense as the German anti-aircraft fire. On the low altitude night incendiary raids Japanese anti-aircraft fire was not very effective, but the smoke, the air turbulence due to heat, the searchlights, and the danger of running into another B-29 in the dark made for a very scary ride.

As for the nighttime fire raids, the jet streams above Japan were the reason why they got started. Typically, the earlier missions were done above 25,000-35,000 feet. At these altitudes, we discovered the jet streams. Before that, nobody even knew that they existed. So, when we turned to go on a target downwind, our aircraft was going about 500-600mph with that tailwind behind us. The Norden bomb sight was not designed to cope with this kind of speed; strategic bombing was impossible. Going east to west these winds slowed the planes ground speed so much the aircraft would be a sitting duck for anti-aircraft fire. North to south or south to north runs caused such excess drift that strategic bombing was impossible. Then, we lowered the altitude and dropped incendiary bombs at night - that worked!

From a high altitude strategic bomber the B-29 became a low level bludgeon of defeat for the Japanese. Before this, we were lucky to even hit the city, let alone a factory. So, General LeMay figured out that these factories were supported by a lot of people who did cottage work. Therefore, we could disrupt their production by burning up their cottage industries all around their factories. Typically, our primary targets were the industrial cities such as Tokyo, Osaka and Nagoya. We also struck some other small industrial cities; for example, one of my last fire raids was against Toyama. We burned 98% of the buildings in that town down.

Pathfinders were assigned for the fire raids and would take off an hour before the rest of us. They would fly to the target city and go to four different locations to mark out a square with fires. In the center of this square would be the area we would bomb. So, when we arrived this would be our target. On most missions, our fire raids left around 1700 hours, we took off at one at a time, flying to Japan at lower altitude. Each B-29 was flying alone to and from the target, managing your own fuel consumption. The navigator, one pilot, flight engineer and radio and radar operators working all night, others dozed until we hit the coast of Japan and flew into what looked like the mouth of hell.

Planes would go in at 5,000-8,000 feet, whatever their assignment was. This was very scary because it was pitch black at night and the bottoms of our B-29s were painted black so the searchlights wouldn't reflect off of them. The radar controls on their anti-aircraft guns were set for higher altitude raids so nothing they shot at us was very affective. The main danger was running into other B-29s in the dark, which did happen. The smoke and the convection from the fires below were so terrifically heavy when we flew into the smoke; everything was just black. Then, the updrafts from the convection of the heat would take the planes and bounce them up, anywhere from 1,500-10,000 feet. Because of this, we would be strapped down very tight and wore goggles, flak jackets and all that stuff.

Updrafts created by the fires burning below were so strong they would bounce the planes hundreds of feet, some times thousands of feet upward when the plane had dropped its bomb load. Our planes were all depressurized and there was one man in our group that had his B-29 flipped upside down. Thankfully, at about 2,500 feet, he was able to right his aircraft. I think, this happened to three different aircraft on these respective fire raids.

You could "smell" the fires in the aircraft because we were not pressurized and when the bomb bay doors opened, the smoke just came in. I've heard guys say that they smelled human flesh, but, I don't think so. I guess flesh may have been burning but, to me, it smelled like a huge wood fire. Initially, anti-aircraft fire was set to high but later on, they set it lower so our planes would get caught in the searchlights. These searchlights would light you up like daylight and when one did this, quickly after, several more found you. We dropped tinfoil, which we called "rope" to confuse the radar down below. It looked just like the tinfoil you would hang on a Christmas tree; they were long streamers. Furthermore, all of our fire bombs had bomblets that had streamers on them to do the same thing (to confuse the radar). We carried 500 pounds of bombs with napalm in them. One B-29 carried enough bombs to start 1,500 fires. As the fires heated up, winds came along and there was a firestorm. The fires were out of control and the Japanese couldn't do anything. Let me tell you, it was scary until you left the target area. After flying all night, we arrived home at dawn. So, we could see a beautiful Pacific sunset and sunrise as well.

The fire raid on Nagoya, 14 May 1945. This is one of the missions flown by Braden and the crew of B-29, "Stripped for Action." (NA)

A perspective of the bomb damage in a typical Japanese city during the war. This is Kobe, the target for a 17 March 1945 mission. (NA)

Feelings while in Combat

As a soldier, you did what you were told to do; it was just a job. When your friends are killed and you were being shot at, you didn't have any particular feelings for people. Furthermore, Japan had started the war and had its civilians making weapons. Later on I read how horrible our bombing missions were, but, I also read that the Japanese had killed over 17 Million people during World War II. They were extremely cruel and did not take prisoners of war. B-29 aviators were immediately condemned to death and some were executed. For example, "Hap" Halloran was stripped naked and chained to the bars of a lion's cage because all of the animals were gone. People would poke at him with bamboo sticks. Hap said that for forty years he couldn't sleep at anybody's house because he had nightmares and would wake up screaming. However, he finally returned to Japan and supposedly found the guy who shot his plane down. This helped him get over it.[4]

Combat did not affect me in any negative way. As I mentioned earlier, I was shot down and ditched on my third mission, rescued by the Navy. Two crew members were killed and another one sent home with a broken back. The survivors were assigned as replacement members on crews that had lost members due to death or being wounded. Along with two other survivors from my original crew, I went on to complete a tour of thirty-five missions to the Japanese homeland with another crew. We landed on Iwo Jima four times either so shot up or so low on fuel that we could not have made it back to Saipan. In all of this, I had my scary moments, but I was never so frightened that I could not do my job. I was never reluctant to fly another mission because I knew when I completed thirty-five, I could go home.

Engagements with Japanese Fighters

On all the high-altitude missions, we had attacks from enemy fighters. In particular, I remember one engagement when we had fighter escorts, P-51s, from Iwo Jima. This mission occurred on 7 April 1945 and our target was Tokyo. On that day, we got jumped by countless fighters, but many of them were shot down. It was a perilous mission. We had one engine shot out and half of the aircraft commander and co-pilot's controls were shot out. One man had the control yoke and the other had the rudder pedals. I believe, there were around 184 bullet holes in our aircraft; there also was a hole in our vertical stabilizer (tail) that looked like you could drive a Toyota through it! That was our first time to go into Iwo Jima, it saved our lives. I don't think we ever could have made it home.

We never saw any night-fighters. We expected to see them but never did. Incidentally, on those night missions we didn't carry any guns except for the ones in our tail. This was done to save weight and carry more bombs. However, there were gunners still on our planes but they served as observers to spot if we got too close to another aircraft, etc.

Towards the end of the war, we didn't have any fighter opposition whatsoever. They were that crippled. We were trying desperately for them to surrender unconditionally. We dropped leaflets which said, "We will be coming to your town on Sunday afternoon, at this (specified time). We urge you to get your government to surrender and to leave your city." But, they didn't leave and we went up and burned it down.

Final Combat Mission

On 8 August 1945, I flew my last mission, my 35th, after they dropped the atomic bomb on Hiroshima. This one was to Kyūshū, the Yawata Iron Works, which was one of the earlier targets. It was a very different route and the entire way back was over water. Therefore, you didn't have that nice chain of the Mariana Islands to check your location (on radar). Our flight engineer was worried about the fuel consumption, which was a big thing in the B-29. For example, on a raid to Tokyo we carried about 7,000 gallons of gas. However, we had to carry more fuel to a place like Kyūshū. The flight engineer was worried that we didn't have enough gas to make it home so he wanted to land on Iwo Jima. Nevertheless, our pilot said that we were going home as this was our last mission. So, we got to Saipan and the engines were still running, but the flight engineer's knuckles were white. By this time, we were all worried as those gauges showed that they were out of gas so we called in an emergency landing. This would be done if you had wounded on board or were really low on fuel; they would let you in the landing pattern early. Just as we were going down the runway, all of the engines quit. Because of this, they had to send out a tractor to haul us off the runway. When I got out of the aircraft, I kissed the ground!

Coming Home

In September 1945, I came home in a war-weary B-29 with three other men, a pilot, aircraft commander and flight engineer. After doing about a two-week leave, I had to report to San Antonio, Texas. A sergeant there asked me, "Would you like to get out?" My reply was, "You bet!" During my military service years, you would get points for things such as the number of missions you flew, time in combat, number of awarded medals, etc. I had a lot of points but didn't know how the process worked. The next thing I knew, "Wham, and with a quick stamp, I was out of the Army!" My last rank was first lieutenant.

Postwar

I WAITED FOR A FEW months to get into the University of Texas. Before the war when I attended college, I studied Aeronautical Engineering. But, I found that my passion was airplanes, not integral calculus; the latter also wasn't interested in me. Since I had already two years of engineering courses, what could I apply them to? Architecture, which typically was a five-year program, but I could transfer all of my prior classes. So, I became an architect and was very fortunate to be successful after working in this field for forty-two years. During my years as an architect, I worked on just about any government agency you can think of; for example, hospitals and criminal justice institutions.

Next, I served on the board for Dallas/Fort Worth International Airport then became the chairman there. Lastly, I was a construction arbitrator with the American Arbitration Association and on their master list of one-hundred construction arbitrators. As far as my family, I got married in April 1946 to Sara Deering. We had three daughters together and now I have eight grandchildren, six boys and two girls. All of my grandchildren are now grown up and doing well.

Fifty years after the war, amazing 20th Air Force veteran David Braden had an opportunity to visit an old friend, B-29 "Fifi." This Superfortress is maintained in flying condition by the Commemorative Air Force. (David Braden Collection)

Reflections

I WAS, AND STILL am, very patriotic. I was incensed by the attack on Pearl Harbor and the Nazi attempt to conquer Europe. When I arrived on Saipan I found that 3000 Japanese civilians, men, women, and children had committed suicide to die for their Emperor rather than surrender; I thought how stupid. Then, I read a diary found on the body of a Japanese soldier on Saipan. In it he regaled on the glory of dying for his Emperor on the attack planned for the following morning. I then began to understand the culture of the people we were fighting. They were absorbed in a culture of death no American could understand.

Then, when crew and friends began be killed around me, I became absorbed with the feeling we had to stop this crazy war, whatever the cost to Japan; I knew we had the vehicles and the weaponry to do it. When it was necessary to kill the entire garrisons of Japanese soldiers to capture Saipan and then, Iwo Jima, I understood what had to be done to stop the war - kill them all. They were crazy and cruel beyond belief. Now, that I am older and a student of the history of the war, I know the Japanese killed 40,000,000 people in the course of their warfare, were guilty of inhuman treatment of their own troops, made slaves of women, raped civilians, worked POWs until they died, and were even guilty of cannibalism in their officer corps. I feel that fighting the Japanese was like stamping evil from the earth. They were defeated by the air raids which wiped out their industrial cities, but their war lords refused to give up and were embarked on a course of genocide for the nation. Only the atomic bombs got the attention of their Emperor who advised surrender, thereby in all probability, saving a million lives - both Japanese and American. As for me, I was lucky, born lucky I guess.

David Braden passed away on 25 April 2015, preceded, by his wife Sara who died on 23 January of that same year.

[1] This bombsight was considered "state of the state" during World War II. The bombardier would input data such as wind direction, airspeed and altitude into this bombsight. Then, its mechanical analog computer would calculate wind drift and provide an accurate aiming point. The Norden bombsight had an internal gyroscope that made it a stable platform for high-altitude bombing. www.nationalmuseum.af.mil

[2] *The United States Strategic Bombing Survey, Effects of Incendiary Bomb Attacks on Japan, A Report of Eight Cities.*

[3] LORAN stands for long range navigation. This system, developed by the United States in World War II, had superb accuracy, range and speed. Navigators utilized intermittent signals from pairs of land-based radio stations to help plot their course. They would use the loran equipment (in their aircraft) to calculate the time difference between the pair of signals (or pulses) then plot their position using a loran chart. www.ibiblio.org

[4] On 27 January 1945, Second Lieutenant Raymond "Hap" Halloran was a B-29 navigator who was shot down during a mission to attack the Nakajima Aircraft plant in Musashino (Tokyo).and became a prisoner of war. He was part of the crew that flew the B-29 named the "Rover Boys Express." Halloran's crew was in the 878th Bomb Squadron, 499th Bomb Group, 73rd Bomb Wing. For more information about this veteran (who is now deceased), please visit: www.haphalloran.com

302 Kōkūtai

The following six chapters (11 – 16) feature the accounts of Japanese Navy veterans who were all aviators and aircraft mechanics serving with the distinguished 302 Kōkūtai. *This unit had the difficult task of protecting the Japanese main island of Honshu and the Kanto region which includes the capital city of Tokyo. They were led by an experienced and career navy officer named Captain Yasuna Kozono. Because of their importance in wartime aviation history, along with the unprecedented access I had to the veterans of this unit, I chose to designate this special section in the book to them.*

Instead of providing the background of this air group and the infamous incident that happened at the end of the war, I will let the surviving 302 Ku *members tell their stories. It has been a distinct privilege and honor to attend this air group's reunion for several years in a row; always being the only non-Japanese person, non-302* Ku *veteran there and befriend these men.*

See Appendix IV, pages 265-267, for rare group photographs of this elite unit.

Chapter 11

"I feel people today really don't understand the 'true' Japanese side of World War II."

Petty Officer 1st Class Mitsuyasu Yamakawa (IJN)

While I lived in Japan and did my research with the surviving WW II Japanese Navy airmen one of my personal "missions" was to locate a veteran who flew the Mitsubishi Raiden *fighter in combat, especially with the well-known 302* Kōkūtai. *The gentleman who several of my buddies recommended was Mitsuyasu Yamakawa. This soft-spoken former fighter pilot took me under his wings and even invited me to come to his unit's reunions where I was able to meet and interview some incredible people. He was one of the best supporters of my work with the veterans and a good friend for many years.*

Early Years

In 1923, I was born in a rural part of Nagasaki Prefecture called Shimabara. I attended a local primary school for eight years. Later on in my teenage years I moved to Nagasaki City and began working in a store there. While living in this city, I remember studying hard. During this time period, our country was fighting in the Sino-Japanese War so a lot a young people were very patriotic; this prompted many individuals to enlist in the military. In 1941, I joined the Imperial Japanese Navy at Sasebo, where I spent four months essentially being indoctrinated into the Navy and learning about serving in the military.

Flight Training to Become a Fighter Pilot

The training airplane for the primary course was a two seat biplane. The (official) name of the plane was the Kugisho Type 93, but its popular nickname was the "Red Dragonfly." This aircraft was adopted by the Navy in 1934. It had 340 horsepower with a maximum speed of 291 kilometers per hour. The first thing I practiced doing was touch and go landings. Typically, a pilot trainee flew for seven to ten hours in the air before he had a chance to "show what he could do." I had to take a test to do this but before taking it, I needed permission from my flight instructor. The second thing I did was flying (with my instructor) and practicing basic aerobatics. After being in the air several times with him, I started to gain confidence in my flying abilities and thought, "Wow, I am becoming a pretty good pilot." I felt this way because my instructor (who was sitting in the seat behind mine) wasn't yelling at me anymore.

Next, I practiced formation flying with two other aircraft. During this type of flying, you really had to keep your eyes open at all times. Initially, we did very basic level formation flying but later it became more difficult with doing additional maneuvers in the air. However, if I could not follow the other aircraft well, the instructor whipped the back of my head. After this, we practiced doing navigation and continued to get more confident in my skills as an aviator. I was even able to look outside and enjoy the beautiful view from the air. After several flights with the instructor, two students flew together on the same plane. In the front seat would be the pilot while in the rear seat, it would be the navigator. During this time, the navigator did his basic navigation training. In the rear seat, a drift-measuring instrument was installed for navigation and the navigator could see the effects of the wind while flying. This instrument really helped the pilot fly in the direction that he wanted to go. The speed of the airplane varied, depending on whether you had a crosswind or tailwind. For example, if there was a crosswind you had to recalculate the drift otherwise you could not reach your destination.

As far as high-altitude flight, you climbed to 5,000 meters (16,404 feet) slowly and this was the Type 93s maximum altitude that it could fly. It took about forty minutes to reach this height so you had to put on an oxygen mask. This was the first time for me to use oxygen and this course lasted five months. By this stage, you had completed almost all of the flight training and were required to do a flying test with typically a *buntaichō* or veteran aviator. They considered "everything" and then decided which aircraft type was best for the flight trainee to fly in the next training course. Next, the students would advance into different courses such as fighters, dive-bombers and land-based bombers. Each type of airplane had its own special course and the trainees were subsequently sent to the appropriate unit.

My secondary course was in fighters and the aircraft I used was the Mitsubishi Type 96 carrier fighter (Allied code name *Claude*). It was a one-man aircraft but very slow. This plane was used in combat before the legendary *Zero* fighter was manufactured. During the four-month training program between January-May of 1943, we started with practicing touch and go landings. Next, I flew my Type 96 very closely to the instructor's aircraft in front of mine; there usually was a distance of twenty or thirty meters (66-98 feet) between us. I had to fly my fighter "exactly" the same way he did. After this, I thought my flying skills became really proficient and we practiced dog-fighting each other in the air. Together, we practiced many different air combat situations. For example, attacking an opponent while in superior position (from above) or being attacked while you were at a lower altitude than your adversary.

Formation flying was usually done with three aircraft and since we used fighters we practiced doing various flight maneuvers such as sharp turns, etc. We practiced a lot and it felt like "real combat" for us, but of course, it wasn't. Finally, the last stage was gunnery. For this training, we used a tow aircraft which had a target sleeve attached by a seventy-five meter cable. We shot at this sleeve from behind at a forty-five degree angle. The bullet heads were painted in four different colors: red, yellow, blue and green. Each pilot trainee had bullets with only one color so we could easily determine how many hits he had. Every aviator received a grade for his gunnery skills and this was an important part of our training. I completed this course in four months.

Next, we flew the *Zero* fighter. This training began in May 1943 and was held at Iwakuni Air Base (Yamaguchi Prefecture). It was almost the same as our previous training except we had more time to practice night flying, touch and goes and acrobatics. We completed the *Zero* fighter training in two months and then were assigned to an operational unit. I went to a land-based unit in the Ōita Prefecture, but the people who got the better grades were assigned to aircraft carriers. Unfortunately, all of the aviators who served on aircraft carriers died during the war. From Ōita, I was assigned to the 253 *Kōkūtai*, which was based in Saipan. As of July 1943, I was officially part of this unit. During the months of July and August, we were training very hard, both day and night, to prepare for the air combat which would be coming soon for us.

Undated picture of a young Mitsuyasu Yamakawa while serving at Tobera Airfield, Rabaul. Note that he is wearing shorts and not a typical flight suit to fly his combat missions. (Mitsuyasu Yamakawa Collection)

*****

Yamakawa-san recalled while on Saipan he visited the town of Garapan with several comrades and admired the natural serenity of island life. He was surprised to see bunches of bananas so cheap, which was unheard of in Japan back then. The two months the young aviator spent on Saipan went by quickly and he was assigned to Rabaul on New Britain Island.

On 1 September 1943, after his operational training was completed, Yamakawa-san and his comrades flew to Rabaul via Truk Atoll. The group, which consisted of about eighty airplanes was led by Lieutenant Commander Harutoshi Okamoto. The flying time to Truk Atoll was four hours then the group refueled there before flying another four hours to their final destination. This was his first long flight as a Navy fighter pilot. After arriving at this Japanese stronghold in the Pacific, the young pilot felt "he had arrived" and was quite excited to be there.

Flying Combat Missions - Fall 1943

I was assigned to Tobera Airfield at Rabaul and was there until January 1944. My primary duties were to do aerial patrol missions for groups of our transport ships. Almost everyday our aviators had to fly these missions that usually lasted around five hours. Sometimes, submarines and enemy bombers attacked our ships. Around this time period, I remember shooting down a B-17 bomber. Other times, I flew to the vicinity around Bougainville with several other aircraft and we did reconnaissance there. I wasn't able to go to Guadalcanal as only "veteran" airmen took part in those missions. Typically, I went as far as the Shortlands and Vella Lavella. For example, during one mission I attacked an enemy airfield at Torokina after it was completed. Since I was flying these combat missions with other aviators that were doing the same thing, I really didn't feel lonely or homesick. I honestly never really felt "scared" because I did this almost on a daily basis. Once, I recall a moment that it was a "close call" and thought I was about to die, but I didn't look back and say it was a scary thing. This is because I was trained this way. During a mission to Bougainville, I strafed some enemy aircraft on the ground and as I flew away from the airfield, my aircraft was buffeted around by anti-aircraft fire. I felt like I was riding a horse but my *Zero* fighter was too slow to get away quickly. To me, that was a close call but I made it back okay.

Feelings About the Skills of his Adversaries

I thought that the Allied airmen were good pilots. For example, while in formation, there was no situation when one aircraft strayed away; they always stuck together. In the beginning of the war, the Japanese Navy fighter pilots were not used to flying against formations and they usually fought individually in dogfights. However, Allied aviators showed that flying formations were effective, so later in the war, Japan did the same kind of training. Nevertheless, most of our pilots were not used to flying against formations so they were easily shot down. Japanese aircraft tended to be very maneuverable because we usually preferred fighting the enemy "one on one." For example, if we engaged four aircraft with two of our planes, we would lose. Therefore, we felt that it was prudent only to engage the enemy on a one to one basis.

Notable Combat Missions – Late 1943

ON 1 NOVEMBER 1943, I attacked LSTs at Bougainville (Torokina area) when they were trying to land.[1] But, this was not the intended purpose of my mission as I was supposed to protect our fleet which had left from Rabaul. My comrades and I could not find our fleet; instead we spotted a large group of enemy LSTs which were trying to land around the Torokina area. We quickly prepared to attack these vessels and lowered our altitude. Our group of aircraft went north, right above the beach and fired on them. Since we were attacking from such a low altitude my fighter was hit numerous times by enemy anti-aircraft guns. Luckily, I was not wounded and my engine appeared to be running fine but started to leak oil. It became hard for me to see as the oil began to cover my windshield so I had to do an emergency landing on Buka Island. I wanted to have the mechanic check the oil leak but he was really reluctant to do so as this airfield was recently attacked. He thought that if my *Zero* fighter was parked on the runway, they would be targeted again. So, I only stayed here for a few hours but he gave me some extra oil to get back to Rabaul.

*****

YAMAKAWA'S NEXT NOTABLE mission took place on 26 December 1943, twenty-five Type 99 kanbaku *and fifty-five* Zero *fighters went to attack ships off* Tsurubu *(Cape Gloucester region) on New Britain Island. The 552* Ku *contributed thirteen Type 99s while the other twelve were from the 582* Ku. *The overall commander of all the* kanbaku *was Lieutenant (junior grade) Tokiyasu Shimizu (of the 582* Ku*). As for the escorting* Zero *fighters, the breakdown per air group is as follows: thirteen from the 201* Ku, *nineteen from the 204* Ku *and the remaining twenty-three aircraft were part of the 253* Ku. *Yamakawa-san was flying in this last group of fighters from the 253* Ku. *The fighters from the 201* Ku *and 204* Ku *were commanded by Ensign Suzuo Itō while the 253 Ku was overseen by Lieutenant Commander Harutoshi Okamoto.*[2]

*****

WE WERE ESCORTING twenty-five dive-bombers (582 *Kōkūtai*) and trying to attack the United States Marines that were landing there. I already flew on another sortie that morning so I was very tired, but my commanding officer needed an extra pilot so he asked me to fly once again. I couldn't even eat because everything was done on such a short notice. I was flying in the last four aircraft formation and in the "tail end Charlie" position (of the entire group) so I figured my *Zero* fighter would be an easy target for the enemy that day.

Our dive-bombers were flying at a 5,000 meter altitude (16,404 feet) while we flew at 5,500 meters (18,404 feet). This allowed our group of *Zero* fighters to protect the Type 99s from above. As we approached the beach area, I saw several enemy vessels in the water below and some of our dive-bombers began their bomb runs. Just at that moment, a group of about ten American P-40 fighters started firing at us but they were still not very close to our aircraft. However, we had to stay with our dive-bombers and protect them at all costs. My *Zero* fighter was attacked from above and behind by a P-40 fighter. It began to start on fire so I had to bail out. After bailing out, I spent about five hours in the water before finally reaching the beach; I ended up sleeping there that night. My face and legs were burned and I had a bullet scar on my left leg. Then, on the following morning, I starting walking towards Tobera Airfield but I found one of our Army bases which was on the same island but east of *Tsurubu*. Several days later, I boarded one of our submarines and was able to return back to my base at Tobera.

As an interesting anecdote, I always flew with seat-pack type parachutes. The parachute itself was the seat cushion so it was hard to imagine flying without one. Without the parachute cushion, you would sit too low in the cockpit so everyone flew with parachutes for all our missions.

*****

Two Zero pilots from the 253 Ku *were lost in action: Petty Officers (both first class), Tadashi Hirai and Chikara Kitaguchi. As mentioned earlier, Yamakawa also bailed out over the water but was later rescued. Japanese records state that one additional* Zero *fighter from this unit made an emergency landing but the pilot's name was not recorded. No airmen from the 201* Ku *were lost on this mission. Lastly, PO1c Tomoharu Honda from the 204* Ku *did not return from this mission.*

Just before entering the combat area, the #1 Chutai *of the 552* Ku *left the rest of the dive-bombers and headed for ships in Borgen Bay. However, enemy fighters were waiting for them at low-altitude and every aircraft in this* chutai *did not return. The 582* Ku *lost four planes including the Type 99 that mission commander Lieutenant (junior grade) Tokiyasu Shimizu was flying in. Losses also were especially heavy for the 552* Ku *with eight of the thirteen strike aircraft being shot down.*

1944 & More Combat for Yamakawa-san

Before my wounds were properly healed, I transferred to the 204 *Kōkūtai*. On 26 January 1944, I flew on a Type 97 Seaplane to Takeshima Island (also known as Eten or Etan Island) in Truk Atoll. Because of the American air attacks by Task Force 58 on 17-18 February, the 204 *Kōkūtai* was devastated in combat. Therefore, several of my comrades and I were sent back to the Japanese mainland. At this point, I joined the 302 *Kōkūtai*, which just formed on 1 March. One week later on the eight of the month, I became a member of this unit as a future pilot of the Mitsubishi *Raiden* fighter. However, there were only ten men (four officers and six petty officers including myself) who flew this fighter at Yokosuka Air Base. The 302 *Kōkūtai* also had a detachment at Kisarazu Air Base, which were doing flight training there using the *Zero* fighters. On 1 May, our group from Yokosuka and the other one from Kisarazu joined together at Atsugi Air Base then our full-scale combat operations began. Our commander, Captain Yasuna Kozono was a very experienced and well-known Navy officer. I was a low-ranking person so I was not in the position to talk with him, but everyone (including myself) respected and trusted him as our leader.

In 1944, a lot of our airmen switched from the *Zero* to the *Raiden* fighter. Even though these fighters were much "different" in flying characteristics, our pilots thought they were the "same." Because of this, there were a lot of flying accidents and countless men perished while training in this aircraft. Many *Raiden* pilots did not like the fighter, but personally, I think that it was a good airplane. However, the truth is, compared to the *Zero* fighter, the *Raiden's* engine was not that reliable. The *Raiden* fighters broke down often, especially due to high oil pressure which would cause the engine to overheat. Nevertheless, after I got used to flying it, I always checked the oil pressure and everything else. After this, I didn't have any troubles. On a few occasions a buddy would be flying next to me and report that he had problems because of oil pressure then the next minute I saw him crash. I thought that this fighter could be "delicate" to fly but once it was in the air, everything was fine.

A group picture taken sometime in April 1944 at Oppama Air Base while these aviators were training on the Mitsubishi Raiden *fighter.*
Front row, left to right: LdfSea Mitsuyasu Yamakawa, LdfSea Sueki Mori, WO Jinan Shikada, Lieutenant (junior grade) Tomiya Miyazaki, Ensign Senri Isozaki, PO1c Takumi Sugitaki and LdfSea Tadashi Sakai. Back row (standing), left to right: PO1c Itaru Shikano, Ensign Tatsuo Yui and CPO Kikuo Nishimori. (Mitsuyasu Yamakawa Collection)

All of these aviators were experienced Raiden *fighter pilots.*
Front row, left to right: PO2c Mitsuyasu Yamakawa and CPO Yoshiō Nakamura. Back row, left to right: WO Jinan Shikada and PO1c Seikichi Izawa. Note, that Yamakawa-san is now a petty officer, second class in rank. (Mitsuyasu Yamakawa Collection)

Sometimes, during take offs at maximum power, our pilots would go off the runway because when the plane was swerving to the right just a little bit, he would suddenly pull back on the stick and it would flip over. During training, I had an accident while flying a *Raiden* fighter, which was just repaired, from Tateyama Air Base (in Chiba) to Atsugi Air Base. Before the flight, the mechanic checked the engine and said that it was perfect but as I was flying over the runway at Atsugi, the engine just stopped! So I tried to land the fighter but it was getting dark. Luckily, our aircraft on the ground were dispersed on the air base but I still hit one of them.

During October 1944, I went to Kasumigaira to test a *Raiden* with a turbosupercharger, but it was very heavy. The center of gravity was towards the front of the airframe. This version of the fighter had a very long nose because it had a normal engine with a turbosupercharger. Typically, in the Navy, we did three-point landings, but this wasn't possible because it was too heavy. Because of this, the aircraft landed nose first. I only tested this model of the *Raiden* one time and it never flew in combat.

The *Raiden* Fighter vs. the B-29 in Combat

THE *RAIDEN* WAS DESIGNATED to be an attack aircraft against the B-29. We developed the way to attack the B-29, which was called *Chokujouhou* (direct up tactic).[3] Our fighters attacked from 1,000 meters (3,281 feet) above the B-29s. Typically, the best way to attack is from the rear, at a 45-degree angle, but the *Raiden* had only four guns. So, it was impossible to use only four guns against the heavily-armed B-29s. After doing extensive research, we discovered the best way to attack was from above at a 90-degree angle. However, at the beginning, you would start from about a 45-degree angle because of the distance between both planes. If you started to dive right above the B-29, by the time you got into your shooting position, you would end up being behind the enemy bomber. We practiced this method in training, but the pilots who were not experienced were not successful. While flying combat missions, I did the *Chokujouhou* several times, but it was very difficult to do. You had to commence your attack at just the right moment. I felt that you could only be "successful" if you were an experienced airman. However, rookie aviators would miss their chance and end up being shot down.

The usual way we fought against the B-29s was by using a tactic called *Koujouhou* (back up tactic). In this method, the most efficient way was to attack them from the rear at a 45-degree angle, from above. But, flying a *Raiden* against a B-29; how can just four guns be affective against the amount of firepower they had? You can try to destroy it, but you probably won't last long. There was also a tactic called the *Zenkahou* (front down tactic), where we attacked from about a 20-30 degree angle, below the B-29, from the front. I used this tactic a lot but the accuracy was very low compared to the tactic from above, where you had more area to hit. About fifty percent of our fighters were unsuccessful and shot down using the *Zenkahou* method. As a unit, we mostly used the *Chokujouhou* method. As for myself, I did the *Chokujouhou* method only when the conditions were ideal. Typically, when the altitude was near 10,000 meters (32,808 feet), I used the *Zenkahou* against the B-29.

We approached in formation, but no formations were used during the actual fighting. The attacks we did were totally dependent on the commander's decision. The commanders were inexperienced so frequently we would miss our chance to do our attacks on the enemy aircraft. We could not go in before the commander, and that may have been part of our disadvantage. When I attacked the B-29s they used to fight back with their countless machine guns, which made a color similar to when you are welding metal. Without a way to shoot them down I might have been scared. Nevertheless, I was concentrating on my attack and couldn't afford to think about anything else at that time. As I dove down on the enemy bombers, I continued to fire my guns while passing through the formation. Then, it took about ten minutes to go up again and do another pass on the bombers. During an attack from above, any B-29 could have shot me down, but from the front it was difficult. If I could not get to a higher altitude in time, I had to make a frontal attack. After I did a frontal attack, I went under the target airplane and dove away. I still believe the best way to attack a B-29 was from above. However, if the first attacking fighter, usually flown by a commanding pilot, was late; all of the aircraft following him missed the B-29 all together. This is also the reason a lot of *Raidens* got shot down. The leader's timing was essential to the success of the attack.

In January 1945, the now experienced Raiden *pilot PO2c Mitsuyasu Yamakawa (right) is explaining the finer points of the* Chokujouhou *attack method the 302* Ku *used to engage B-29s to PO1c Masaru Tajima (left). They are going over this method on the rooftop of the command post. (Mitsuyasu Yamakawa Collection)*

Below: The Zenkahou *attack method which Yamakawa-san said he preferred to use when engaging B-29s that were flying at an altitude of around 10,000 meters (32,808 feet). (Michael Claringbould)*

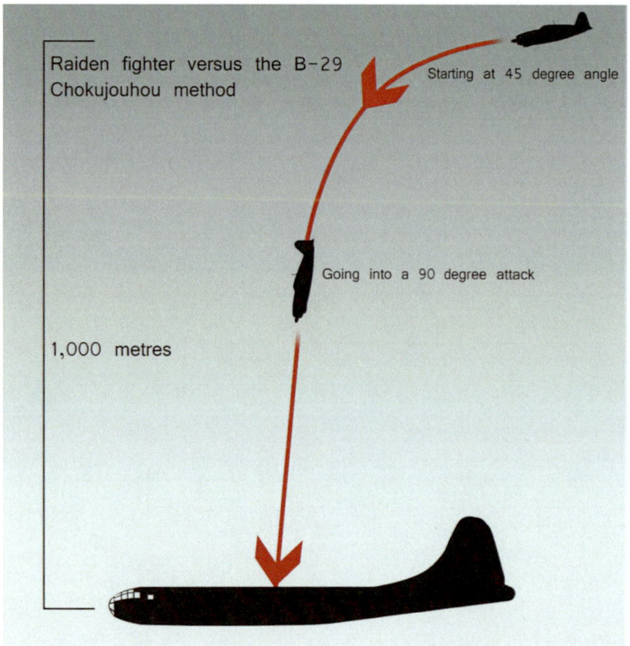

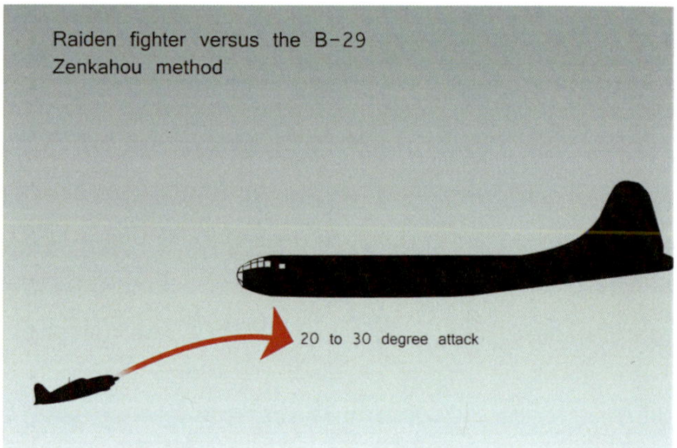

Left: The Chokujouhou *attack method which was utilized by the 302* Ku *to try to shoot down the well-defended B-29 Superfortress bombers. (Michael Claringbould)*

I THOUGHT THAT IT was almost impossible to shoot-down a B-29. If the bomber flew around 5,000 meters (16,404 feet), it would have been a good match. However, when we flew at a height of 10,000 meters (32,808 feet) or higher, the B-29 was faster than our fighters. This is because it had turbosuperchargers. At this altitude, our aircraft engines did not perform well and they lost speed. Therefore, many of our pilots were shot down. Flying at such a high altitude took a lot of effort by a pilot just to keep his fighter in the air and put extra stress on the aircrafts engines. We had very few chances to attack these bombers because if the *Raiden* pilot saw it, usually he could not catch the aircraft because of its speed. The B-29s even had self-sealing fuel tanks that seemed to be "bullet proof." At that time, Japan did not have this feature in our aircraft. As for the engines, the B-29 had four of them so even if one engine was damaged it could still fly. Unless we shot at some critical points such as the rudder or elevators, we simply couldn't shoot them down.

The Use of Oblique Cannons in the 302 *Ku*

WHILE SERVING AS A commander of the 251 *Kōkūtai*, at Rabaul (1942-1943), Captain Kozono installed the obliquely-mounted cannons on the *Gekkō* fighters. This modification to these fighters produced good results and they shot down several B-17 bombers. Later, after Kozono transferred to the 302 *Kōkūtai*, he had oblique cannons installed on the *Gekkōs* and other aircraft. Our commander really spent a lot of time and energy on this effort to modify our aircraft and had great expectations for them. At first, these modified fighters were successful in shooting down some enemy aircraft, but after March 1945 when the war was coming to an end, they were becoming less successful. These cannons also were installed on our *Zero* fighters and a night-fighter group was formed. However, I think they were not that successful in combat.

Daily life with the 302 *Kōkūtai*

OUR SUMMER AND WINTER schedules were different. In the summer, at 0600, we got up and then ate breakfast. After this, we went to the control tower for a briefing. In front of the headquarters other non-flying personnel gathered together in the square. Our schedule depended on the weather; for example, when it was cloudy we could not practice gunnery so we did formation training. At the briefing we looked at a blackboard that had each person listed on it. According to this, we did our training. It showed several groups, but which group flew first varied depended on the situation. For a mission, the pilots' names were already on the blackboard and listed their aircraft positions. For example, this aviator would be flying the number one aircraft, etc. This blackboard even listed each formation of four aircraft and the position of each pilot in the group. It was rare for every group to fly. At Atsugi Air Base, the total amount of aircraft was usually less then fifty and typically around twenty percent of them were not flyable due to damage or that they were being repaired. Furthermore, there were only a handful of *Raiden* pilots because it was more difficult to fly than the *Zero* fighter.

A briefing prior to flying a mission at Atsugi Air Base. (S. Bon Group)

In the morning, we were always ready to go and took off when we received the warning. At that time radar was not that sophisticated so it was rarely used. Instead, the incoming enemy aircraft were spotted visually. For example, when

somebody saw seventy or eighty B-29s fly past Hachijō Island they would calculate how long it would take to reach Mt. Fuji area. It took us about forty minutes to fly to 10,000 meters (32,808 feet) so everything had to be calculated in advance. There were observing posts that used binoculars and they contacted Yokosuka Headquarters who in turn informed us at Atsugi Air Base. After we received this information, the *Raiden* pilots took off and flew according to it. When a formation of B-29s was coming in, we all took off and went after them. But, if there were only one or two B-29s came to do reconnaissance, only a few *Raiden* fighters went up. The maximum flying time for this fighter was only two hours, but if a belly tank was used it could fly for four hours.

Aircraft mechanics played a vital role in keeping the planes of the 302 Ku *in the air. Yamakawa-san (right) is pictured with one of them, Yamaguchi-san (later changed his family name to Nishide). (Mitsuyasu Yamakawa Collection)*

302 Ku Raiden *fighters during flight operations at Atsugi. (S. Bon Group)*

The Spring of 1945

In April and May of the last year of the war, I recall seeing the P-51 Mustangs coming over Japan. When the P-51s came by themselves they tried to shot down our planes, but when they came, as escorts, with the B-29s, these fighters stayed with the bombers to protect them. At that time, the B-29s flew at 5,000 to 6,000 meters (16,404-19,685 feet) while the P-51s flew at 7,000 meters (22,966 feet). The Mustangs looked similar to our Army's Kawasaki Ki-61 *Hien* fighters so at first we didn't know that they were the enemy. While flying at low altitudes, this fighter was much faster than the *Raiden*.

*****

It was during the spring of this year when the Battle of Okinawa was really at a climax. The Allies increased their pressure of the mainland of Japan by conducting ever increasing air attacks on military facilities on Kyūshū. Because of this, a detachment created from the Raiden *units of the 302, 332 and 352* kōkūtais *transferred to Kanoya Air Base. Yamakawa-san was part of this defense unit called the* Tatsumaki Butai *(or Tornado Group). According to his records, the men from the 302* Ku *became a part of this group on 23 April 1945 and they served together for around three weeks. The overall commander of this group was Lieutenant Commander Kiichirō Nishihata and the 302* Ku *aviators were led by Lieutenant Kushichirō Yamada. Both men were from the 302* Kōkūtai. *Yamakawa-san thought that they fought well against the incoming B-29s which were typically flying at a fairly low altitude such as 6,000-7,000 meters (19,685-22,966 feet). However, while this hybrid defense unit had some success, their casualties and damaged aircraft began to take its toll. They also had trouble getting additional supplies to continue flying combat missions from Kanoya Air Base. In the end, sometime in May 1945, Yamakawa-san and the surviving 302* Ku *airmen returned back to Atsugi Air Base.*[4]

The End of War & the Unit's "Revolt"

The thing that I remember most is when the war ended. When I talk about the 302 *Kōkūtai* I can't avoid talking about our "revolt." Nevertheless there might be some mistakes in my account of this event because it happened half a century ago. I think there are a lot of books about the 302 *Kōkūtai* revolt, but some parts are inaccurate. On 15 August 1945, the day the war ended, there was an announcement and we were told to gather at the airfield. We heard the Emperor's message together, but it was not very clear and difficult to hear. Nevertheless, it was obvious that we had lost the war. I personally didn't feel like listening to this announcement at all. Some people thought that it was useless to listen to this message about "losing" the war. After this, we were all very busy for a few days. For example, some people started to clean up their quarters, and throw away their weapons and military related items. I didn't have much time to eat for two days because it was difficult to organize everything. I had to say goodbye to the landlord of the apartment building where I lived. Furthermore, I burned everything that would be useless after the war.

In the middle of all the chaos after the defeat of Japan, there was a revolt led by Commander Kozono. Originally, Atsugi Air Base was not the type of place to have any trouble like this but everything changed when a junior Army officer visited our base.

After this, some people got excited. I felt like I didn't have the nerve to continue fighting and really didn't know what to do. I think, however, everybody knew that this refusal to surrender by our group would not change the outcome of the war. Nobody thought about the possibility of winning again, but we couldn't just quit fighting. We were not taught to surrender and wanted to continue fighting until we all died. Some flyers were dropped from the air over the Kantō region to let people know that they wanted to continue fighting.[5]

On 17-18 August, it was okay to still fly, but by the 20th, this wasn't possible anymore. Air was taken out of my *Raiden's* tires so I was not able to fly anymore and participate in the revolt. One of the senior mechanics, Maintenance Lieutenant Commander Minoru Yoshino, ordered his men to disassemble our aircraft and take off the propellers. I also recall that he put bicycles under the *Zeros* to prevent them from taking off.

My superior officer Captain Junrō Terumura didn't want the *Raiden* pilots to continue the war and do anything drastic. Some of the men did not want to continue fighting and just wanted to go home. But, I wasn't that kind of person. I wanted to still do something and not just quit and go home. That's why I wanted to ask Lieutenant (junior grade) Sadaaki Akamatsu about what we should do now. I remember talking to him about this while he was buck naked. I felt he was a true leader and the kind of person everyone could trust so I wanted his opinion. This distinguished Navy veteran believed that our base would eventually be taken over by the occupation forces of the United States and this was first time that I heard this. Akamatsu-san wanted some people to go with him to the mountains to escape. I wanted to join his group and continue the fighting. The plan was to meet him with two other men. The one fellow was from Okinawa and the other guy was from Kagoshima. Akamatsu-san didn't have anything with him such as machine guns because he hid them somewhere. We were supposed to meet him at midnight by the gate but later we heard that Akamatsu-san and his group actually left around 2300 hours. The three of us waited for two hours or so but never saw him and his group again.

On 26 August, General Douglas MacArthur was supposed to be coming to our air base so I was told to leave as soon as possible. I received train fare, but didn't know where to go. But in the end, I went back to the Nagasaki Prefecture with the two other men (that I mentioned earlier). The one guy stayed with me for ten days before going back to his hometown. The other fellow lived with me for two years because Okinawa was occupied by enemy forces after the war was over.

Postwar

I STAYED IN MY hometown for a few more years before moving to Kure City then I went to Yokohama. At the latter place, I worked at a gas station. Postwar Japan had a new aviation industry and I got a job as a test pilot for Toyo Airplane Industrial Corporation, which was located in Fujisawa City. However, I left after two years because this company couldn't pay me anymore.

In February 1955, I joined the Japan Marine Self-Defense Force and worked there until joining Japan Air Lines (JAL). Initially, many of the senior JAL pilots were former American servicemen and acted very harshly towards me so I got pretty stressed out. I worked for JAL until retiring in 1984. During my postwar years, I flew the DC-4, DC-6, Convair 880, 727, DC-8 and YS-11. My total flying time (both military and postwar years) is around 18,000 hours in the air.

*****

YAMAKAWA-SAN AND HIS wife, Kikue, had two sons together, Kazuhiko and Kiyonari.

Reflections

I FEEL PEOPLE TODAY really don't understand the "true" Japanese side of World War II. A lot of the WW II veterans want to share their war stories but it is difficult. I feel this is because Japan really cannot tell our side of the story because we are always concerned that this will upset the Chinese and Korean people.

This kind-hearted veteran had a stroke at the age of eighty years old and passed away seven years later in 2010.

Even after all these years, Yamakawa-san recalled his flying techniques used during the war. Here, he is describing flying the Raiden *to the author. (Ron Werneth)*

[1] LST stands for landing ship tank. This type of allied vessel carried troops, tanks, cargo or vehicles directly to a beachhead during landings. They were not used with docks or piers.

[2] All of the background information including the number of aircraft flown, losses and other details were taken directly from the 26 December 1943 *kōdōchōsho* for the 201, 204, 253, 552 and 582 *kōkūtais*.

[3] Properly translating and doing follow-up research on the techniques utilized by the *Raiden* pilots against the B-29 bombers was done with the additional assistance of several historians in Japan including Takehiko Shibata, Minoru Kamada and Ikuo Shimizu. I also had a detailed translation done by US-based Koji Ikenaga who kindly helped with this difficult research.

[4] Yamakawa-san shared this interesting story about his time at Kanoya Air Base during his interview with Masaya Kurita and Toshiko Kurokawa when they visited his home.

[5] This region of the mainland of Japan is on the island of Honshu. It includes the major cities of Tokyo and Yokohama. The area includes seven prefectures: Chiba, Gunma, Kanagawa, Saitama, Tochigi and Tokyo.

Chapter 12

"We have to do everything in our power to prevent needless wars from happening again."

Maintenance Lieutenant (junior grade) Shigeō Nakagawa (IJN)

DURING MY THREE-YEAR stay in Japan, I wanted to interview some World War II veterans who served as Navy aircraft mechanics because their stories are usually forgotten. Many people tend to focus on the aviators, maybe because their stories are more glamorous. However, it was the mechanics and other ground crew personnel that had to keep the aircraft flyable and ready for combat.

One of the mechanics I was able to meet was Shigeō Nakagawa, a member of the 302 Kōkūtai, *which participated in home defense missions towards the end of the war. Nakagawa-san did maintenance work on the advanced Mitsubishi* Raiden *fighter. We met together at the 302* Kōkūtai *Memorial Service, which is held every spring at a small shrine in Kanagawa Prefecture to remember the members of their air group who died in the war. Nakagawa-san is mentally sharp and fit for his age. He seemed to really enjoy talking about his wartime experiences with a younger person.*

Nakagawa's Early Days

I WAS BORN ON 20 August 1923 in Shinjō City, Yamagata Prefecture. My parents were farmers and I was the fourth son in our family. I graduated from a local elementary school and junior high school. After this I moved on to a national technical school called Yonezawa High Industrial School, which is currently called Yamagata University. There I majored in engineering and graduated in September 1943.

Training to be an Aircraft Mechanic

ON 23 SEPTEMBER 1943, I entered Ōihama *Kōkūtai* and the opening ceremony was later held on 5 October. I entered as a 7th Term aviation mechanic/engineering preparatory student. Since there was a wartime draft, I knew that I would have to join the Army or the Navy. Because of my engineering background and interest in the Navy, I applied to be an aircraft mechanic. The Navy accepted me and I felt that it was very important to do something for my nation because we were at war.

During the opening ceremony, the Navy officer who spoke to us said that the purpose of our training was the following: to promote military spirit, to develop our character and get the basic training required to be an aircraft mechanic and finally, to develop the skills needed to be an officer-level maintenance man. The "basic" training included the following: spirit education, discipline inspection, physical education and land fighting. We had classes on such topics as aircraft maintenance including fuselage repair, engines, flight instruments and radios. The students studied how to repair, modify and remanufacture aircraft parts. Our group also learned about the prevention of aircraft accidents and what was needed to either build a new air base or move one to another location.

After entering the Navy until 14 November, which was about one and a half months, we had training that focused on mental and physical growth. As far as classroom studies, the men studied about the proper conduct of a Navy officer in both wartime and peacetime situations. Our group did those classes then after three weeks we started to study aviation engineering.

Daily schedule of my training class according to Mr. A, who was my classmate.

0555 – An announcement from the PA system stated, "Five minutes before wake up. Five minutes before wake up."

0600 – "All members wake up. All members wake up." Then, everyone got up or jumped off their hammocks. After this the hammock was stored and everybody went to the washing room. There, people were divided into groups and lined up at their pre-determined spots for training. It took five minutes to do all these things. After inspection, the students did stretching exercises. These exercises were not the slow stretching; instead, they were rhythmic exercises

0755 – We heard the announcement stating, "Five minutes before lining up for the lesson," then we lined up in front of a drill platform.

0800 – Another announcement blared out, "Lesson Start," then the instructor on duty said what the lesson would be for that particular day. After that, the students would go to their classrooms. We did our classroom studies in the mornings.

1200 – After finishing our classes, we ate lunch. The students had free time until the lesson in the afternoon so we organized our own personal belongings or did the laundry.

1300 – The physical drills for basic training were conducted. These included the following: flag signals, shooting firearms and how to use a sword. The men did stretching exercises, running, judo, kendo, land fighting and cutter boat training. We also practiced giving orders and sometimes sang military songs.

1745 – The drills were finished so the students could have dinner and take baths.

1830 – We reviewed our classroom lessons of the day and got a preview of the following days curriculum.

2030 – The trainees stopped studying and went back to the barracks and got ready for bed. After this, there was an announcement to lower our hammocks and we changed into our pajamas.

2055 – Another announcement stated, "Five minutes before inspection." Then, all of the men went into their hammocks and waited for our nightly inspection.

2100 – We heard an announcement that there was an inspection and then a trumpet was sounded. After the inspection was over, we had some free time until the lights were turned off.

Beginning on 15 November, we began our technical classes. From this time we started to have more time in the classroom using technical textbooks and the daily schedule was similar to the one above. In addition, we had land-fighting classes and learned military songs. The trainees also studied wire knotting and flag signals. In the technical classes, the men used Mitsubishi G3M2 Type 96 land-attack airplanes (Allied code name *Nell*) and *Zero* fighters for maintenance practice. Furthermore, the students worked on Kinsei and Kasei motors. We practiced disassembling; assembling, test running and doing typical maintenance work on these engines.

Left: 1 January 1944, Shigeō Nakagawa during his early days in the Imperial Japanese Navy. (Shigeō Nakagawa Collection)

Right: 31 May 1944, just after Nakagawa-san was promoted to the rank of ensign. (Shigeō Nakagawa Collection)

The first official class outing for Nakagawa's class was to visit the historic battleship Mikasa *(October 1943). It was flagship of Admiral Heihachirō Tōgō during the Russo-Japanese War of 1904–1905. Today, this over 100-year old warship is preserved as a museum in Yokosuka. (Shigeō Nakagawa Collection)*

302 *Kōkūtai* days

Starting in late May, we had a maintenance course for about four months at Mitsubishi Heavy Industries. Our group learned about the Mitsubishi *Raiden* fighter and propeller-maintenance at Sumitomo Metal Industries. On 10 July, our assignments were decided and five days later, I graduated. Immediately after this, I was assigned to the 302 *Kōkūtai*.

The mechanics spent most of our time doing maintenance work on the *Raiden* fighters. I was a member of the maintenance group and my duties were to supervise all the work being done on our aircraft. I did this job until the end of the war. My main duty was to work as a mechanic so I didn't do anything else. I just focused on doing my daily maintenance work and I really didn't seem to have a schedule to follow. The most important thing for me, as a mechanic, was to increase the number of airplanes, which were flyable. First, I got up and ate breakfast. Then, I did some more work on our aircraft and ate lunch. Normally, I ate lunch at a command post inside the airfield. After this, I did some more maintenance work and had dinner. Sometimes, I had additional work to do at night as well. Lastly, I took a bath and went to bed. Every day was like this.

I don't remember what I did in particular when I didn't work. In 1944, we had a lot of flight training by the pilots so we, mechanics, also were very busy. Nevertheless, maybe until the fall, we were able to go out and I had some free time at night. After this, however, we had to have our aircraft ready to fly at night because of possible enemy air attacks. But in the beginning, it was not busy at all. In the evening, we really didn't have to do anything special. Later, because of these air attacks, our airmen had to fly very often to intercept the enemy aircraft and the mechanics had a lot more maintenance work to prepare our fighters.

In order to protect our aircraft from enemy air attacks, I remember that we moved fighters to a pine grove at the edge of the airfield to cover them with pine branches for camouflage. I recall covering the fighters with blue nets and then placing branches on the nets to "hide them" from attacking enemy aircraft.

Above: Nakagawa-san checking out the cockpit of his favorite fighter, the Mitsubishi Raiden. *(Shigeō Nakagawa Collection)*

Right: The young Shigeō Nakagawa during his service with the famed 302 Ku. *(Shigeō Nakagawa Collection)*

I think, the mechanics had great relationships with the pilots of our group. As far as I remember, there was no special treatment for anyone who worked on the same base. Personally, I didn't have any personal problems with any of our aviators. Instead, I focused on doing maintenance work so I didn't have time to talk with the pilots.

Maintaining the *Raiden* Fighter

I HAVE A STRONG "attachment" towards the Mitsubishi *Raiden* fighter because it was really the only aircraft I worked on (while in the Navy). However, compared to our other wartime fighters such as the Mitsubishi *Zero*, it was a difficult aircraft to maintain in good operating condition. After each mission, we did maintenance work on our *Raiden* fighters such as checking the oil pressure and changing the spark plugs on the engines. As for the airframe, the landing gears and control stick were inspected and nuts and bolts were tightened, etc. Since this fighter needed constant maintenance, it was difficult to maintain in "perfect" flying condition. Frequently, *Raidens* had oil pressure problems and sometimes roller bearings in the engines got burned, but they were later improved. It also had a lot of lubricant leaks, which caused the aircraft's windshield to get very dirty and poor visibility for our pilots. The fighters landing gears were also not that strong and we often had problems with them. We didn't have any difficulty getting spare parts or tools because we had the important duty of protecting the Japanese capital, Tokyo, from air attack.

Early models of the *Raiden* were armed with two 7.7mm machine guns (in the fuselage) and two wing-mounted 20mm cannons. It also could carry two 30-kilogram bombs. The most numerous produced version of this aircraft was the Model 21, which had four wing-mounted 20mm cannons. This version of the fighter also could carry two 30 kilogram bombs or two 60-kilogram bombs.

The *Raiden* fighter unit proceeded to the Kanoya Air Base (in Kyūshū) from 23 April to 16 May. Although most members of the unit came back to the Atsugi Air Base on that day (16 May), I came back to the base one week later because I had to take care of some unfinished duties.

A nice close up image of the netting used to help camouflage some of the 302 Ku *on the ground at Atsugi Air Base. (Shigeō Nakagawa Collection)*

Nakagawa's original caption for this wartime photo, "Maintenance work continues." (Shigeō Nakagawa Collection)

Another wartime picture of aircraft maintenance work. Here, Nakagawa-san is lending a hand to help repair an engine on a 302 Ku *Gekkō fighter at Atsugi Air Base. His caption, "There is a lot of hard work behind any victories during combat." (Shigeō Nakagawa Collection)*

Above: *When not repairing of flying their aircraft, members of the 302* Ku *managed to get in some games of volleyball for relaxation and exercise. (Shigeō Nakagawa Collection)*

Right: *302* Ku Raiden *fighters preparing to take off at Atsugi Air Base. (S. Bon Group)*

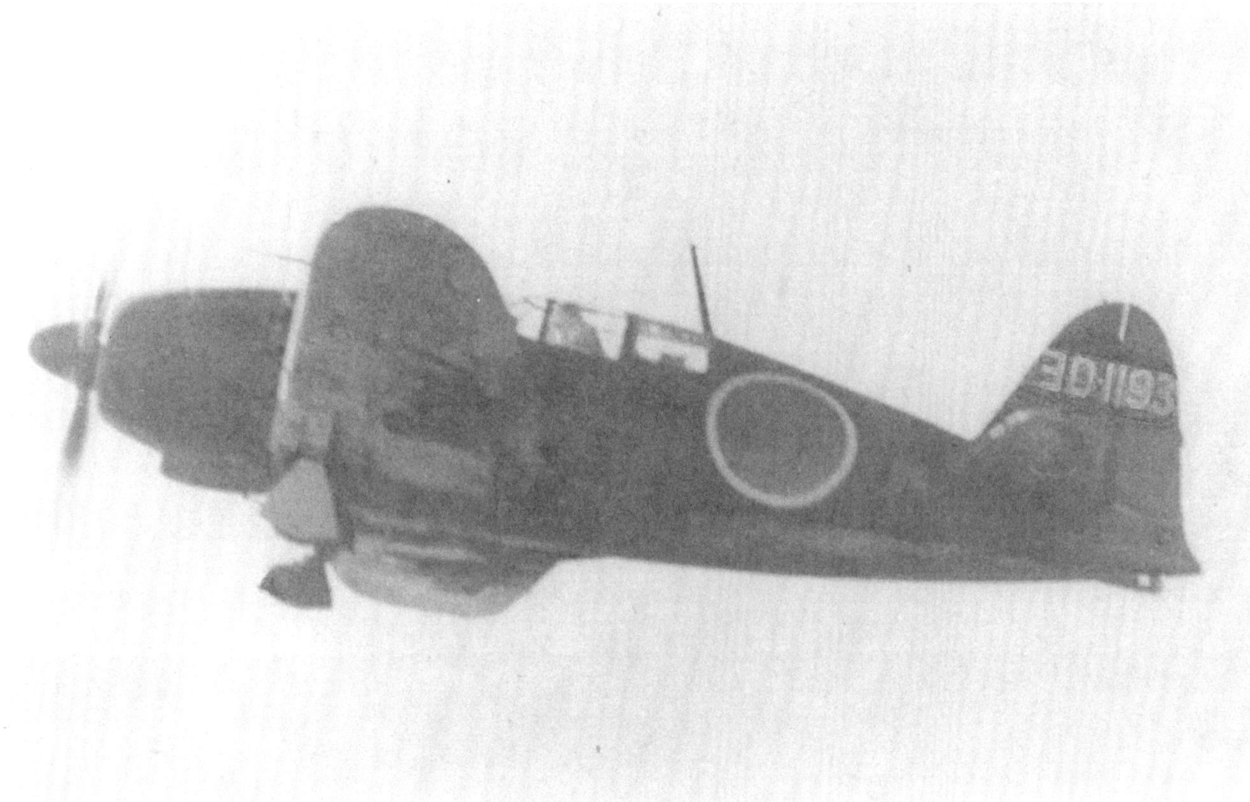

Gear up on a 302 Ku Raiden *as it climbs after take off. (S. Bon Group)*

The End of the War

At the end of the war I was still a member of the 302 *Kōkūtai*. Everyone gathered at the airfield and listened to the radio. I really can't say how my other comrades felt after hearing the news but from their expressions, I thought, everyone looked a little delusional and really didn't know what to do. Personally, it was hard for me to think about the future or what I should do next. After hearing the Emperor's message, there was talk that we should continue fighting until we died and many of my comrades wanted to do this. I believe, during this time, I still continued to work on our aircraft and really wasn't directly involved in the events that occurred during 15-23 August 1945. I think our commander was ill and other men from our air group persuaded him that we should continue fighting even after the Emperor's message. I went with some other men to the mountains in the Nagano region, where we wanted to make our "last stand." The members in my group included *Zero* night-fighter pilots Fujiwara-san and Katsu-san, Motobayashi-san, a *Zero* night-fighter mechanic, and some other petty officers. There were about ten of us in our group but our truck broke down during our journey and we gave up on our idea.

Postwar

At the end of August 1945, I was demobilized and sent home to Yamagata Prefecture. I moved to Kawasaki in February 1946 and lived there until July 1948 so I didn't have much of a chance to go to Tokyo. However, I heard in Tokyo, many people lost their houses due to the fire-bombing during the war and some of them lived in tinplate shacks. Food such as rice, vegetables, fish, etc. was distributed by the government, but it wasn't enough. The amount of rations was less than 50% of the amount we received before the war and the rice was only enough for one meal. So, people tried to get food from local farmers. For example, some residents exchanged precious metals or clothes for something to eat, but sometimes the police confiscated their food. In the cities the "black market" was rampant and a lot of people bought food and daily commodities there. This actually saved starving citizens at the time. There were many orphaned children because they lost their loved ones in the war and nobody was able to take care of them.

After the war I went to my hometown, but early the following year I came to Kawasaki. I bothered my relatives and they took care of me. I visited many companies to get a job, but nobody wanted to hire me. I was able to live in my relative's house until I finally joined Toyo Communications in May 1946. Three years later, I moved to Tokyo and still live here today. Right after the war ended, there wasn't much to do and my paycheck always came late. Furthermore, every time I got paid, my earnings were very little.

When I joined this firm, I moved into a company-owned apartment, which was only four and a half *tatami*. Food rations (such as rice) were distributed by the government, but it wasn't enough so I went to farm-houses to buy yams, etc. I sold half of them and survived on the other half. Since I didn't have "civilian" clothes, I took the insignia off my old navy uniforms and wore them every day. A lot of people did this. I was one of the luckier ones because I was still single and didn't get wounded during the war.

I worked for this firm until July 1948 then joined the Japan Otis Elevator Corp. After joining this new company, I did mostly engineering related jobs. In the 1950s, Japan went into a recovery period and we had a lot more elevator construction projects. I used to work on my holidays and had a lot of overtime work along with all-night construction jobs. This continued on and on. I worked for this company until August 1983. After this, I joined OC Engineering Company and retired in March 1996. I have lived in Tokyo since August 1948 and enjoy being retired. I got married after the war, but we have no children. I enjoy playing golf and the Japanese game called "Go" to relax.

Fellow 302 Ku *aircraft mechanic buddies and lifelong friends, (left) Yoshiō Oshitani and Shigeō Nakagawa during a postwar reunion. (Ron Werneth)*

Reflections

In my case, I was lucky that I didn't have to fight overseas during the war and my assignment was on the Japanese mainland. I believe it was my duty as a Japanese citizen to fight for our nation against the unwarranted pressure given to Japan by the United States. Out of a total of 941 graduates from my training group, the 7th Term Mechanical Reserve Class, 130 men died in the war.

Although Japan is still in a mild recession, we still have it "good" because there is no trouble getting any material goods or food. I believe that these days are "good days" because it is a free society now. Nevertheless, at the same time, it's too relaxed in Japan now and people have lost their identity as Japanese. People are losing their morals such as misconduct by police, politicians and women who are prostitutes. There also are many egocentric people without common sense. Today I hear unimaginable crimes for pre-war standards happening, for example, child abuse by parents, atrocious crimes by young boys and an increase of murders done by people against members of their own families. I think all of the Japanese people need to work together and resolve these issues.

Many younger people don't understand or learn about the sacrifices my generation made during World War II. However, we, the veterans, have a duty to send the message about the facts of the war in order to avoid the meaningless tragedy again.

In the last war not only combatants died, but also many innocent people perished. We have to do everything in our power to prevent needless wars from happening again. Sadly, even after this useless killing, wars continue all over the world today. Because of this, it is hard to have hope for the future. I really can't stop these negative feelings in my heart either. In general, more analysis needs to be done about prior wars. For example, what were the causes of past conflicts and how they can be prevented from reoccurring in the future? I hope that one day people of all nations can learn to respect and value the different cultures, traditions and history of their counterparts around the world. Then, maybe some further wars can be prevented through mutual understanding, knowledge and respect for the different groups of people throughout the world.

Chapter 13

"I never thought about whether I would survive or not."

Maintenance Petty Officer 1st Class Yoshiō Ōshitani (IJN)

As I began to seriously delve into researching and reading about World War II, I always was curious about the experience of a typical Japanese Navy aircraft mechanic who served on an island base such as in the Solomons. Unfortunately, as I carefully checked with my veteran friends and attended reunions, I discovered that there were few survivors left to talk with. Luckily, I was able to meet a gentleman named Yoshiō Ōshitani who was very fortunate to survive several dangerous experiences during the war. He not only served on a remote Pacific base, but also on an aircraft carrier, and lastly, with the elite 302 Kōkūtai, *in the defense of the Japanese mainland. I was able to meet Ōshitani-san on several occasions and always found him to speak in a very matter-of-fact way but happy to share his stories with me.*

Life Before the Navy

I was born on 1 April 1923 in the Shiga Prefecture. During my teenage years, I worked in Osaka, as an apprentice, for a small trading company. This firm exported materials to Germany but because of the war, it was difficult to do this type of work, so my company went bankrupt.

The Start of the Pacific War

On 1 September 1941, I joined the Imperial Japanese Navy. At Maizuru, I did a three-month boot camp then graduated on 27 November of that same year. Then, on 4 December, I was assigned to the 19th *Kōkūtai*, stationed in the Ebijie Island (part of the Marshall Islands). Four days later (in Japanese time), Pearl Harbor was attacked and Japan went to war with the United States and its allies. I ended up staying there for six months before being sent back to Japan for additional training.

Next, on 1 June 1942, I became an 82nd Term student at the maintenance school (in Sagamino, part of the Kanagawa Prefecture). In November of that year, I completed this course, and, the, on 15 December, I transferred to the carrier *Ryūhō*. On board this vessel I served as a mechanic for the *Zero* fighter. Our busiest time was when the pilots did landing practice. This is because we tried to do everything within three minutes including unhooking the aircraft from the landing wire and then putting the fighter on the elevator. Everything had to be manually done. On 12 June 1943, the *Ryūhō* joined the 2nd Carrier Division. Three days later we left Yokosuka then arrived on 21 June at Truk Atoll. From there, our carrier air unit joined the 204 *Kōkūtai* and went to Rabaul, on New Britain Island. I stayed at Rabaul for over twenty days before going via the destroyer *Matsukaze*, on 5 July, to our base at Buin (Bougainville Island).

An image taken at Rabaul; very similar to the type of work Oshitani-san and his squadron-mates did in this tropical environment. (Ron Werneth)

1944-dated portrait of Yoshiō Ōshitani when he was maintenance petty officer second class. (Yoshiō Ōshitani)

Difficult Days on Buin with the 204 *Ku*

LIFE WAS VERY HARD at Buin. After being there for almost two months I had lost a lot of weight mostly due to the continuous bombing raids on our base. While I was there I drank only boiled water and the people who didn't do this got very sick from dysentery. It was vital to boil water for three minutes before drinking it. I learned this the hard way because earlier in my Navy days with the 19th *Ku*, I was hospitalized for eleven days and literally stayed in the bathroom for much of that time. So, I didn't make this same mistake again.

Just about every morning, enemy aircraft which were flying in a large formation, would bomb our base. We tried to make trenches but it was difficult because they would fill up quickly from groundwater. I recall when twenty of our men were killed when a time-delayed bomb later exploded.

I think there were about twenty aircraft at Buin, but typically; only ten or less, were in flyable condition at any given time. This was due to either mechanical issues or simply that we didn't have the replacement parts to repair them. It was very difficult to even get parts to our base and so much was destroyed from enemy bombing attacks. Each aircraft would normally have about five of us mechanics to maintain it. Rookie maintenance men would usually take care of a pilot's fighter that was about the same (in experience levels).

On most days, I woke up at 0500 in the morning and we had to get the aircraft out of the jungle. Then, I would crank the inertia-starter of the engine on a *Zero* and it could take about sixty minutes to warm up (in the South Pacific). This was very important so the engine wouldn't get damaged. When I was stationed in Japan, warming up an engine would be much easier (using a battery) and take just five or six minutes. Then, a handful of the fighters (usually 3-5) took off for their sorties.

If there was an air battle, our pilots would usually return in about thirty minutes. However, if they flew an escort-type mission with Type 1 *Rikkō* bombers from Rabaul, the fighters would return between three-five hours later. Furthermore, if just enemy fighters came, we would engage them. But, if heavy bombers attacked us, our fighters really couldn't do much. Around 1900, in the evening, after the aircraft returned we had to put them back into the jungle. This job took about two hours to do because you had to cut down the jungle vines then put sheets of iron on the ground so we could move the fighters more easily. In those days, the enemy had control of the air so if we pushed our aircraft into the jungle, they would have a better chance of not being hit during a bombing attack.

When I was not taking care of our aircraft, I could relax a bit. However, it was important to still stay nearby them because I could still be needed to prepare a plane for flight. During that time, I tried to look for my food to eat, but it was very difficult. From the jungle, we found some eatable things such as leaves from a pepper plant. I would boil these leaves in sea water which would give it some taste from the salt and provide vitamins for me. I also ate some Japanese canned goods.

During my over two-month stay at Buin, there were a lot of American attacks on our base. I recall during one of these attacks, my buddy and I were trying to escape while an aircraft was strafing us from behind. Suddenly, my comrade got hit in the stomach and smoke literally came from a hole in this area (of his body). He tried to say something to me but I couldn't understand it then fell over and died. If one of our men had perished, we took his body to the beach and covered it with palms. However, it was even difficult cremate bodies because the enemy could see the smoke from the fire and attack us. Nevertheless, we always tried to do this so the ashes of the deceased men could be returned to their families in Japan.

I was able to interact with the aviators from my group. We all went on the same ship together to Buin. Because of our "seniority system", if the pilots were the same age as I, it was easy to become friends. However, if he was "senior" (age/experience). All together, about twenty of us went to Buin together, but only about seven men returned back to Japan.

Feelings While in Combat

When we went to Buin, it was a "struggle." We were probably some of the last men to go there and survive. Everything was difficult and I didn't think about much else. We did this for sake of the Emperor and Japan. I never thought about whether I would survive or not. People, who "worried" too much about their survival, wouldn't make it through combat.

Military Life after Buin

On 16 September 1943, I left our Buin base. My journey home took me through Buka Island, Rabaul and, lastly, Truk Atoll before arriving in Iwakuni, Japan. A tense part of this trip occurred during the first leg off of Buka Island. Since our ship couldn't move during the day, I had to stay on this island for a bit. During that time, we endured an intense thirty-minute air attack but luckily, I didn't get wounded.

After arriving in Japan, I had a seven-day vacation before getting back on board the carrier *Ryūhō*. On 19 October, my ship left Saiki (Ōita Prefecture), and sailed to our base at Seletar in Singapore. We stayed there for two months, which was like "heaven" to me. There were no enemy air attacks and we did training for about two hours a day. The rest of the time was just like a vacation for us. After this our next stops were Manila, Tarakan and Palau.

Then, on 22 December, we went to Truk Atoll to ferry aircraft, pilots and support personnel (such as me). All of us and the planes were unloaded on the island, which had several airfields.

We expected enemy air attacks and I was there during the big raids on the island. A little before 0600, an air raid alarm sounded and we went to our airfield, but it was too late. There was a sea of fire and a chain reaction of subsequent explosions, so we couldn't even get close to our aircraft. Without even having any breakfast, I fled to take cover near a small hill. Anchored in our big military harbor were numerous vessels, both from the Imperial Navy and commercial ones, everything from battleships to hospital ships. However, I think about eighty percent of them were sunk.

My unit, the 204 *Kōkūtai*, which was commanded by Captain Takeō Shibata, suffered heavy losses. Only six or seven fighter pilots were able to survive because of the rest of them were rookies and had come from Japan to do more training. It was so difficult for him to see only his men getting shot down one after another, while their American opponents lived to fight another day.

I stayed on Truk Atoll until sometime in March 1944, when I received orders to go back to Japan for additional aircraft mechanic training class. I left there alone via a cruiser and went to Saipan Island. I had to wait on this island for about twelve days until I could go home. This was just a few months before Saipan was destroyed by the enemy. However, when I was there it was very serene; almost like there wasn't a war going on.

I am sure if I hadn't left there, I would have died on that island later on. As I sailed back to Japan, I felt our ship could have been struck by an enemy torpedo at any moment. Because of this I volunteered to watch for submarines so I couldn't sleep for three days. The captain was so impressed by this and complimented me. But, personally, I thought it was better than sleeping on the lower deck of the ship, where I probably would perish if we were hit by a torpedo.

302 *Ku* Days

In April 1944, I went to Sagamino again for six months of advanced training. I was a 64th term student and studied the Kasei engine, which was used on the Mitsubishi J2M *Raiden* fighters. After I graduated in September (of that same year), I transferred, in October, to the 302 *Kōkūtai*, based at Atsugi, Japan.

My duties with this unit were similar to my previous assignment with the 204 *Ku*. I was an aircraft mechanic for the *Raiden* fighter. After getting up in the morning, my first task was to help move the aircraft which were scattered around the airfield in order to try to protect them from enemy air attacks. Next, I had to warm up the engine on the fighter and this took a lot more time to do in the winter months. Starting this respective aircraft motor was pretty difficult to do. Many inexperienced mechanics couldn't start it on the first try. If this was the case, you had to remove and then replace twenty-eight spark plugs before you could try starting the engine again. I think, at just about any given time, only about fifty percent of our *Raidens* were flyable and could take part in their assigned missions. As an anecdote, I believe a lot of our pilots couldn't survive if their aircraft were damaged in aerial combat because the Kasei engines made the *Raidens* nose heavy. This made it difficult to bail out in time because a damaged *Raiden* who lose altitude and head towards the ground very rapidly.

In April 1945, I was part of a detachment from our unit that was sent to Kanoya Air Base (in Kyūshū). Just ten minutes after I landed at Kanoya, we were attacked by two enemy aircraft. Immediately, because of my past experiences in the South Pacific, I recognized the sound made by non-Japanese airplanes and, luckily, found a small hole to take cover in. However, some other people weren't so fortunate. I believe seven or eight men were killed in a barracks while they were having a farewell party for another buddy.

Then, in June, I returned back to Atsugi Air Base and stayed there until the war ended. Once the war ended, my commanding officer (for aircraft mechanics) told me to go home until he notified me to return, but this never happened. My final rank in the Japanese Imperial Navy was Maintenance Petty Officer 1st Class.

Raiden *fighters lined up for flight operations at Atsugi Air Base. (S. Bon Group)*

A 302 Ku Raiden *fighter takes off from Atsugi for its mission. (S. Bon Group)*

The 302 Ku utilized several different types of aircraft for combat operations at Atsugi Air Base. Fighters included the Zero, Raiden *and* Gekkō. *They also flew the* Suisei *and* Ginga *for their respective missions. This superb panoramic view was taken at Atsugi Air Base. (S. Bon Group)*

Postwar

I GOT MARRIED AND we had one son and two daughters. My wife is now deceased. As for my postwar job, I worked as an electrician for almost fifty years before retiring.

Reflections

DURING THE WAR, I *knew* Japan would ultimately lose. This is because we didn't have many ships or aircraft left anymore, which really started after our disastrous loss during the Battle of Midway. In our wartime years even a small elementary school would have a special place designated to put the Emperor's portrait. If there was a festival, the school principal, who wore a tuxedo with white gloves, would remove this special picture so it could be placed in the auditorium and people could pay their respects. At that time, dying for the Emperor was on the minds of everyone.

Yoshiō Oshitani during one of the author's interviews with him. He was a good friend of many years and assisted with the research required for this book. (Ron Werneth)

Chapter 14

"Our education system taught us that the Emperor was sacred and we should sacrifice our lives for him."

Lieutenant Minoru Toyonaga (IJN)

While I lived in Japan, I was fortunate to meet many amazing veterans from the 302 Kōkūtai *when I attended their annual reunion held in Yamato City (Kanagawa Prefecture). While attending these reunions, I was very keen to locate a veteran who served as a night fighter pilot that flew combat missions against the mighty B-29 Superfortress bombers. Minoru Toyonaga was introduced to me and he was one of the few Japanese aviators I met who spoke English well. After our initial meeting, he became a good friend and I was fortunate to interview him several times. We were honored to have Toyonaga-san attend our wedding in Tokyo.*

During one of our meetings together, he showed me his wartime photo album which was filled with countless pictures of some of the men from his pilot training class. As I flipped through the pages of this album, the ex-Navy pilot pointed at one picture after another and sadly grimaced. He said, "This man died in the Aleutians. He died at Rabaul This other guy got shot down over Taiwan." It was the same story over and over as we looked at these photographs. Furthermore, Toyonaga-san was just one of a handful aviators who flew nighttime combat missions over the Japanese mainland and lived to tell about it.

Toyonaga's Early Days

I was born in 1921. When I was a student, there was a club for people interested in flying. Quite frequently, many of these same people joined the Navy Reserve. I graduated from Yokohama Technical College and since I was interested flying I joined the Navy. I had two elder brothers and two elder sisters in my family. One of my brothers graduated from college as an engineer and later he did this same duty in the Navy.

On 20 January 1942, I joined the Imperial Japanese Navy at Yokosuka. When I joined, they asked me if I was interested in flying and I was. Then, I was given a flying aptitude test and passed it. I also had a medical check up and then was transferred into the pilot training program. In those days, at the age of twenty years old, you would get drafted into the military. However, if you were a college student, you could postpone this requirement for another two years, but once you graduated, you had to join the military. Normally, most people joined the Army. The Navy had a test to join the Reserve (*Kaigun Yobi Kōkū-dan*). Officers in the Imperial Japanese Navy should graduate from the Naval Academy. But, *Yobi* officers could be graduates of colleges that had maritime-related curriculum.

I was a member of the 10th Term Reserve Group. However, before doing any flying, we did our basic training at Iwakuni Air Base. There, we did a lot of physical training such as rowing (in boats) and studied some core Navy curriculum, for example, signal communication using flags. During my flight training, I felt every instructor was excellent. Typically, I recall the flying time on these practice flights was fifteen or twenty minutes in the air. My instructor was sitting behind me (in a two-place) training aircraft. We practiced doing take offs and landings three times. After twelve hours of flight training, I flew solo for the first time. Since the instructor wasn't flying with me, my aircraft felt lighter in the air.

Left: An undated picture of Toyonaga-san during his early days as a naval aviator. (Minoru Toyonaga Collection)

Below: A rare private photo of legendary Navy ace, Lieutenant (junior grade) Sachiō Endō, when he was a flight instructor at Kasumigaura from May-December 1942. It was during this time that Endō-san taught a young Minoru Toyonaga some flying skills. Their paths would later cross again while serving in the 302 Ku *together. (Minoru Toyonaga Collection)*

As an anecdote, well-known Japanese fighter ace, Commander Sachiō Endō was one of my flight instructors when I was learning how to fly. Later in the war, when we were both members of the 302 *Ku*, he was killed in combat. So, there was a strange connection between us. I thought Endō-san was an accomplished aviator and flew his aircraft with great prowess in the air. I felt he was very easygoing and Endō-san didn't yell at me. Furthermore, he didn't worry about small details.

Off to the Frontlines in the Pacific

After my training was done, I served as a flight instructor for the *Yokaren* pilot trainees. Next, I transferred to the 321 *Ku* (*Tobi Butai* or "Hawk" group) and went to Katori Air Base (Chiba Prefecture). In May 1944, we went to Tinian Island (part of the Marianas Islands) where I

In late 1944, Lieutenant (junior grade) Sachiō Endō (in the foreground) strutting away from his Gekkō fighter during his days with the 302 Ku. This amazing aviator flew for fifteen years in the Imperial Japanese Navy until his death, in action, on 14 January 1945. During his long flying career he saw carrier duty on board the Akagi *and* Sōryū *along with assignments both in central China and Rabaul. Lieutenant (junior grade) Sachiō Endō was nicknamed "B-29* Gekitsuiō*" (King of B-29 Killers). Counting only B-29s, he was credited with shooting down eight and damaged eight more. After his death, Endō-san received a double-promotion to the rank of commander. (Ron Werneth)*

did anti-submarine patrols. We did these patrols and reconnaissance work, which could get scary. During a typical mission, we didn't fly just to one direction and return back; we often changed directions several times. So, I often worried if my navigator would bring us back safely without getting lost. Next, I went to Yap Island and commanded a detachment there. After this, I went to Peleliu Island. During this time period, the majority of our unit was annihilated by carrier-based aircraft from the United States Navy fleet. Then, American forces captured Peleliu Island so we escaped to Davao (Mindanao, Philippines). Because of these heavy losses, we were reorganized as the Fighter 901 *Hikōtai*, which was part of the 153 *Ku.*[1]

In April 1944, at Katori Air Base (Chiba, Japan), two Gekkō *fighters from the 321* Ku *are making preparations to fly to the frontlines in the Marianas Islands. (Minoru Toyonaga Collection)*

While we were based at Davao, no radar was used. During the daytime, some Army men worked as aircraft spotters, and at night, searchlights were used. As for our *Gekkōs*, they had no radar systems in them. Later in the war, our aircraft had radar but it didn't work too well. Therefore, while I was at Davao, we were not able to shoot down many enemy aircraft.

On 1 September 1944, B-24s came to attack us. At that time, only B-24s came without any escorts. The following day, the same type of bombers came again to attack us but this time they were escorted by P-38 fighters. We took off to engage them with four *Gekkō* fighters and two *Zeros*. I was flying one of these *Gekkōs* on this mission. Suddenly, I was chased by three P-38s but I was not really in a position to fight against them because my aircraft had only the forward-firing cannons, mounted obliquely in the fuselage, which wouldn't be useful against fighters. So, I flew as low to the water as possible and tried to escape. Our aircraft was hit from behind and shells started zip through the fuselage. During this attack, my navigator WO Mitsuō Tōda was killed. I was hit and wounded in the leg so I thought it was best to ditch the *Gekkō* in the water before getting shot down. I landed in the water about 100 meters (328 feet) from the beach (near Davao) and an Army boat came to rescue me. After the war, I learned that the P-38 pilot who shot me down was First Lieutenant Robert T. Crow Jr. of the 35th Air Group.[2]

*****

According to the official Japanese account of this 2 September 1944 air battle, as Toyonaga-san stated, at 0900, four Gekkō *fighters and two* Zeros *took off from Davao. Then, at 1030, an estimated thirty B-24s started to bomb runs and soon after, the Japanese side tried to engage them until twenty P-38s appeared and chased them off.*[3]

Return to Japan

After I was wounded, I went back to Japan on a hospital ship to recover. When I returned home, I went to a Navy hospital in Yokohama. On 1 December 1944, I was promoted to the rank of lieutenant. After I got better, in January 1945, I became a member of the 210 *Ku*. I was a member of this air group until early May 1945. Our duty was to protect the airspace over Nagoya and I remember engaging B-29 bombers on three occasions. The return fire from the B-29s guns was very intense. Furthermore, earlier in the war, the speed difference between bombers and fighters was substantial. But, this changed with the B-29s and on many occasions, we couldn't even engage them.

302 *Ku* Days

While I was a member of the 302 *Ku*, there was no "typical" day. Earlier in the war, we were able to do training during the daytime. However, by this stage in the war, we couldn't afford to do this. For example, during the day, if enemy aircraft were attacking from the south, we fled somewhere up north to escape them. Sometimes, we stayed at Atsugi Air Base and tried to camouflage our parked aircraft using tree branches, covers and whatever else we could find. We dispersed the aircraft around the base so they were not easy targets. On other occasions, we intercepted the enemy aircraft. We didn't have any trouble getting enough food but gasoline was a different story. Because we were so short on fuel, we couldn't even fly training missions. It had to be conserved for flying only combat sorties.

Previously unpublished photograph of Lieutenant Minoru Toyonaga during his 302 Ku *service, discovered in a private photo archives during the research of this book. (S. Bon Group)*

Lieutenant Minoru Toyonaga awaiting a mission at Atsugi Air Base. (Minoru Toyonaga Collection)

Nighttime Combat Flying

I believe you could fly combat missions at night if you had around 800 flying hours. However, towards the end of the war, we didn't have many experienced aviators so novice pilots sometimes flew these missions. Some aviators took drugs to help them see better at night. We called them "night vision" pills. Personally, I never took them though. If you had some flying experience, it wasn't that hard to fly at night. Our runways were not lined with lights on the sides but we did have two small lights to help the pilot get the correct gliding angle. Furthermore, even without a lot of lights at night, you could still make out the shadows of the buildings and homes near the airport which was helpful while landing.

Fighting the B-29s

Initially, our *Gekkōs* had two fuselage-mounted oblique 20mm cannons firing forward and upward along with two 20mm cannons shooting forward but downward. While flying above the B-29s we tried to swoop down and fire at them using our downward-mounted cannons. Then, we dove under the target aircraft and tried to hit its belly using our upper cannons. However, our downward-mounted cannons mounted on our *Gekkōs* were not that effective in combat so eventually they were removed. I think, the *Gekkō* was an excellent aircraft to use as a night fighter. During my time with the 302 *Ku*, we used several different aircraft as night fighters including the *Zero, Saiun, Ginga* and *Gekkō*. Nevertheless, the *Gekkōs* seemed to get the best results. I thought that it was affective against the B-29s at night, but during the day, it was very difficult to fight against them.

When I flew with the 302 *Ku*, our aviators would receive a message from one of our radar installations in places such as Iwo Jima (before it was captured) or Izu Ōshima which informed us of an incoming raid by enemy aircraft and we took off to engage them. Typically, we waited at different altitudes, and then we would target an enemy bomber which was illuminated in the searchlights because at that moment, it would difficult for them to see us. During our dive down to attack, we gained a lot of speed while descending and we could target them very accurately. We tried to hit the wing roots and fuel tanks on the B-29s. Looking back on my combat missions, my daytime attacks on the B-29s were the scariest. My aircraft was literally "showered with enemy bullets." However, at that time, I didn't have time to feel scared.

I recall shooting down a B-29 bomber during the nighttime/early morning hours of 25-26 May 1945. Just before midnight (about 2200-2300 hours), probably ten of our aircraft took off to engage some B-29s to the west of Atsugi. The enemy bombers came in one by one and continuously. While I was waiting at 6,000 meters (19,865 feet), I spotted a B-29 below me, which was illuminated in the searchlights. I descended to attack this bomber and while flying southwest of Tokyo, I scored hits on it near the wing root area and fire came out. This B-29 started to lose altitude and flew towards Sagami Bay. After shooting down this bomber, we damaged a second one but could not shoot it down. During the summer months (after Tokyo was already burnt to the ground from firebombing), we didn't have many chances to fight against enemy aircraft since it wasn't a target anymore.

A 302 Ku Gekkō *fighter taking off at Atsugi Air Base. Note the aviator with the white flag in the foreground; this duty was often done by junior airmen. (S. Bon Group)*

The Final Days of World War II

During these last days of the war, we heard rumors that American troops were doing landings at Sagami Bay. We received information that large groups of enemy ships would conduct landing operations here. As I took off (by myself) to do reconnaissance of this area, I thought I wouldn't survive this mission. Fortunately, it turned out that luminescent organisms had lit up the water and they were misidentified as enemy ships so I felt very lucky.

Commander Kozono was well-trusted by the men in our unit. He believed that Japan was a nation of God. When we gathered to hear the Emperor's announcement to the end the war, I recall him saying this. "We will not surrender. It was the Emperor who spoke on this announcement but it wasn't his true feelings. People around him made the Emperor do this. We will fight until the end. You can either go or stay with me and fight." However, nobody left because everyone trusted his judgment. Nevertheless, at that time, Commander Kozono was suffering from a bout of malaria, had a fever and really wasn't coherent. In the evening of 15 August 1944, I was assigned to drop flyers from the air over the northern Kantō area (such as Maebashi and Ōmiya cities) to military installations such as airfields urging that we should continue fighting. Then, on the following day, the same thing was done in the Tohoku and Kansai regions.

Because Prince Nobuhito Takamatsu was concerned about what was happening at Atsugi Air Base, he called the base and spoke with Lieutenant Yutaka Sugawara, Vice-Commander of our group. The next day, Sugawara-san went to headquarters to discuss the situation. He understood everything and told us that we should stop fighting at once. In general, officers seemed to differ in their feelings to continue fighting compared to the enlisted men. The latter felt very strongly that we should not surrender. Sugawara-san then said to the men, "If you don't want to follow my orders, why don't you just slit my throat?" After this, our aircraft were disarmed and the tires were deflated. Some of the men who wanted to continue fighting flew their aircraft to Sayama and Kodama Air Bases.

Most of the men in our unit supported Sugawara's decision and we disbanded on 23-24 August. Six days later, General MacArthur came to Atsugi Air Base. Because we lost the war, I didn't want to go home. Instead of going straight home, four or five of us went to Fukushima together because one of the guys had a home there. I ended up staying at his house for about a week then I went home. Since I was too embarrassed to return home during the day, I went back at night.

Postwar

After the war, Toyonaga-san worked for thirty-one years for Fuji Electric then he retired. The ex-aviator moved to Tokyo in 1946 and got married in 1956. They had no children together. He enjoyed numerous trips overseas for business and visiting aerospace museums such as the National Air and Space Museum (NASM). During his visit to NASM, Toyonaga-san had an opportunity to visit with an old friend, a Nakajima Gekkō, *which is the only complete survivor of this wartime aircraft. He also was interviewed by the Tamiya Model Company when they were researching the* Gekkō *fighter to better understand the aircraft for production of their model of it.*

Reflections

In those days, the Japanese people believed that we had to protect our homeland and fight for the sake of our country. Once the war started, we had to contribute to our nation. We never questioned if this was right or wrong.

Our education system taught us that the Emperor was sacred and we should sacrifice our lives for him. Almost every Japanese person thought that they should fight for Japan, except if you were a communist. Today, this is probably hard to believe but we were educated this way.

Picture taken during a 2002 reunion of the 302 Kōkutai. *All of the veterans pictured were former* Gekkō *pilots. Left to right: Longtime friend of the author and translator Keigo "Hammer" Nakahama, the author, Minoru Toyonaga, Kozo Ōhkawa and former aviator (name unknown). (Ron Werneth)*

I feel that younger people today don't try very hard to understand the war. Personally, I try to share with others how tragic war can be but I believe younger people can care less about it. In general, I think they are not interested in learning about our history. If things continue like this, history may repeat itself. This tragedy shouldn't be repeated again and each person should take the time to learn our history and understand why it happened.

I recall talking to a lady whose boyfriend was serving on the carrier *Kitty Hawk* in the Persian Gulf. She said that he may not make it back safely. I told her back in my time this is what happened in the war and almost every Japanese person experienced this. For example, children would have no idea on where their father was and they prayed for his safe return home. I believe, if you have never experienced the effects of war first-hand, many people have simply no idea what it is really like.

[1] The 321 *Ku* officially disbanded on 10 July 1944. *Japanese Naval Aces and Fighter Units in World War II*, 181.

[2] Confirmed this information with historian Jim Sawruk using Frank Olynyk's *USAAF (Pacific Theater) Credits For The Destruction Of Enemy Aircraft In Air-To-Air Combat World War 2*.

[3] 153 *Kōkūtai Kōdōchōsho*, 02 September 1944, via Minoru Kamada.

Chapter 15

"We were taught if you served in the military, dying was a good thing."

Petty Officer First Class Kozo Ōhkawa (IJN)

WHILE I WAS RESEARCHING the 302 Kōkūtai *and attending the group's annual reunions, I met Kozo Ōhkawa. This former* Gekkō *fighter pilot was one of the last living veterans to have fought against the B-29 bombers during their missions over the Japanese mainland in the last year of the war. Ōhkawa-san was very articulate, physically fit and described well his dangerous aerial engagements that took place over six decades ago. My translator "Hammer" and I took a lengthy, multi-transfer train journey and visited him in his longtime home in a rural part of the Gunma Prefecture.*

Early Years

Petty Officer First Class Kozo Ōhkawa in his flight gear. (Kozo Ōhkawa Collection)

I WAS BORN ON 8 December 1925 in Sakae Town, Gunma Prefecture. I had nine siblings in my family, but two of them died very young. I joined the Imperial Navy for several reasons. First, we were taught that we should contribute to our nation. Second, it was very important to respect our parents. Lastly, my father was a farmer and we were very poor. Since our social ranking (in society) was low, I wanted to move up in living status. As an anecdote, in Japan's postwar years, my younger brother joined the Self-Defense Force and became a high-ranking officer, but he is now retired.

Joining the Navy

BACK THEN (during the wartime years), I knew that I would be drafted in the military anyway so I joined the Navy. In our military class system, when a person joined the military service was very important. Therefore, if someone joined just two months after I did that person was considered to be much lower in status than me. On the other hand, if a man joined the Navy two years earlier than I did, he would be considered a "god."

On 1 April 1943, I joined this branch of service. I did my six-month basic training at Iwakuni Air Base. Then, I went to Yatabe Air Base for four months of flight training. After this, in May 1944, I went to Usa Air Base to dive-bombing training with the Type 99 *kanbaku*. Next, in November of that same year, I went to Kisarazu Air Base before ultimately joining the 302 *Kōkūtai* in early 1945.

Flying the *Gekkō* Fighter

I THOUGHT THE *GEKKŌ* fighter was reliable and didn't have too many mechanical issues. It was powerful and a newer design (of aircraft). However, the *Gekkō* was the only plane that I flew in combat so I had to trust it. I felt a lot more relaxed flying this aircraft (compared to a Type 99) because it had two engines so if one went out, I could still return home safely. Maximum altitude was 9,700 meters (31,824 feet) and its top speed was similar to the *Zero* fighter, 150 knots maximum.

Flying a single-engine aircraft was much easier for me. This was the first-time for me to fly a twin-engine plane. Normally, I just used my hands and legs to fly without too much effort. However, for flying the *Gekkō*, you had to use a lever to increase the speed. Initially, this was difficult for me to do, but at that time, I was much younger so after one month of flying this aircraft, it was much easier to fly.

302 *Kōkūtai* Memories

TRAINING FOR ATTACK methods against the B-29s was always done on the ground with models. In my opinion, there was no good way to attack this bomber. Our *Gekkō* fighters could be fitted with a piece of armor plate that would be mounted between the pilot and observer it because this plate would protect only my

302 Ku flight operations at Atsugi Air Base. Note the Gekkō fighters in the foreground with some other types of aircraft behind them including the Suisei *and* Ginga *bombers. (S. Bon Group)*

cockpit area and not the observer. Furthermore, if you used this armor plate, it would throw off the accuracy of our compass on the aircraft. Lastly, in regards to using or not using a parachute, we didn't have any official policy on this; it would be up to each man to decide what to do.

Regarding the flying partners (observers) that I flew with; they were frequently switched for lower-ranking pilots such as me. We also had to switch aircraft and didn't have the power to choose which one we would fly. All of this was pre-determined on our schedule roster and typically the "best" aircraft were already taken by the higher-ranking aviators of our unit.

Of course, the 302 Ku *had to fly its combat missions in all types of weather. In this image, you can see snow removal being done around some* Gekkō *fighters in front of the 2nd* Hikōtai *Command Post at Atsugi Air Base. (S. Bon Group)*

On 14 January 1945, a B-29 was flying at very high-altitude, about 10,000 meters (32.808 feet) so it was difficult for the *Gekkō* fighter to climb this high. So when we reached around 8,000 meters (26,247 feet), it was difficult to climb higher. This bomber was very fast and hard to fight against, especially with an aircraft that had an obliquely-mounted fuselage gun. Typically, we tried to go under the enemy aircraft and do our attack. However, the B-29 was very quick and we had to be "waiting" for them so it was hard to be successful. Our survival rate was very good and thought this was a "waste" but orders were orders. This incident occurred after a group of B-29s completed their bombing attack and they reduced altitude but increased their speed to return back to base.

I was flying above a cloud when a B-29 was descending in the sky. I increased my speed and trying to go underneath the enemy bomber to attack it, but no bullets came out of my guns; they were empty. The proper thing to do was test fire your guns before making an attack but I was never trained this way. If a "conventional" attack didn't work, we also could ram the enemy with our aircraft. Since I was flying above Toyohashi Air Base, I landed and a mechanic loaded our guns with ammunition. Just as I was about to leave, he said that one of my comrades was shot down and I had to try to find him. I flew to the Atsumi Peninsula and it turned out to be pilot Commander Sachiō Endō, along with his observer Lieutenant Shinya Ozaki, who were shot down. Later, they were cremated by us.

I don't recall the exact date, but on one occasion, seventeen 302 *Ku* aircraft were sent to Itami (Osaka City area). While I stationed here, B-29s attacked Nagoya and were on the way back home, flying at a lower altitude, maybe 4,000 meters (13,123 feet). From above, I attacked one of the bombers in this group but I am not sure if my attack was effective or not. The *Gekkō* was designed to be flown at a night and really not that effective during the daytime.

When B-29s came to attack, they were typically escorted by fighters. When this occurred, our aircraft really couldn't do much to stop them so either we tried to hide our planes or escape from the area and go to places like Maebashi. Sometimes, we dispersed our aircraft around the air base and tried to camouflage them under the trees.

Looking back, on 12 April 1945, I think my scariest mission happened but I wasn't scared at that time. On this mission, my aircraft and three other *Gekkō* fighters were in the air over the Kōfu City area (Yamanashi Prefecture). My flying partner was CPO Minoru Higashi (from Ōita, Kyūshū). We were in the second-position of our formation, following the plane flown by my senior officer, Ueno-san. Once we climbed above the clouds, we were suddenly jumped by several enemy aircraft (probably P-51s). They started to shooting at us from behind and I saw the bullets hit my *Gekkō*. I tried to bank my aircraft left and right to avoid this intense enemy attack. Because of my previous experience as a dive-bomber pilot, I tried to escape by diving away quickly. Ueno-san was able to go to Kōfu safely, but I went along the Kisa River and eventually came to the Mt. Fuji area, near the Pacific Ocean. I crash-landed on the beach near Sagami Bay and the impact was terrible. One of the engines and the landing gears were torn off the *Gekkō* as a result of the crash. There were holes everywhere on the fighter and it had lots of damage.

The End of the War

On 15 August 1945, at Atsugi Air Base, we were told to gather in a hangar and listen to the Emperor's announcement. However, his message wasn't that audible and the Japanese he spoke was not the typical conversational language we spoke. So, it was hard to understand what he was saying. I knew that we were in a difficult situation but I thought that he was trying to encourage us to put forth our best effort. Before this, we never heard anything that indicated Japan was going to lose the war so I never thought we would surrender. However, that night, somebody said that we had lost the war. After I understood what happened, I panicked and began to cry. At that time, I didn't even know why I was crying. However, because of Captain Kozono, I tried to prepare myself to continue fighting. Nevertheless, as the days passed, I became an emotional roller coaster for me. The *Gekkō* pilots were not involved in the revolt, even though there was a tense atmosphere and effort by some of our unit to continue fighting. In fact, some of our higher ranking officers severed the elevators on our aircraft so they couldn't fly anymore.

The main objective of my unit was to fight the B-29 bombers. While we were in training, I was supposed to go to the frontlines at Davao, but didn't go. If I went there, I would have certainly died. Initially, I was told to go to Davao as a part of a group of three aircraft. After picking the *Gekkōs* up at the Nakajima Aircraft Company's Koizumi plant, we did some flight-testing of the aircraft. However, one of our veteran aviators broke my aircraft so only two *Gekkōs* ending up going. So, I stayed at Atsugi Air Base and because of the constant American air attacks, they couldn't afford to send one person to Davao. Because of this, I survived the war.

Even though I engaged B-29s just three times, I flew numerous missions to protect such cities as Tokyo and Yokohama. I believe my flight hours are around 700-800, but I can't be 100% certain on this. Unfortunately, I was instructed to burn my logbook and not bring it home. My final rank in the Imperial Japanese Navy was petty officer first class. After the war, I took over my father's business, a local supermarket. Today, my son runs our family business. In my postwar years, I got married and we have one son and two daughters.

Reflections

During the time that I was born in Japan, essentially the only things that I knew were either taught to me in school or what my parents told me. We were educated to hate Japan's enemies such as the United States. This lasted until I joined the Navy at eighteen years of age. We were taught if you served in the military, dying was a good thing. Of course, this was scary, but our education helped us get over this instinctive feeling. Everyone wanted to fight for our nation but when I look back, the Japanese education system was like this because of economical reasons. When you study our modern-day history, it typically begins with Pearl Harbor but there are many things that happened before this. After the Meiji Restoration, Japan had cheaper labor costs. This wasn't good for Britain, Europe or the United States. At that time, Japan's main industry was textiles and we could manufacture them much cheaper than these countries. Therefore, they gave us pressure and we had to react to this.

After the author's visit in 2003, former Imperial Japanese Navy fighter pilot Kozo Ōhkawa outside of his residence. (Ron Werneth)

Chapter 16

"I felt relieved that I could fulfill the promise that I made to Prince Takamatsu."

Maintenance Lieutenant Commander Minoru Yoshino (IJN)

While researching the 302 Ku *and trying to learn as much as I could about this famous air group, I knew that it was vital to understand what happened at the end of the war through the men who were there during the infamous "Atsugi revolt." As I met several 302* Ku *veterans, I heard the name of Maintenance Lieutenant Commander Minoru Yoshino repeatedly and that he was the man who helped hand-over Atsugi Air Base to American Army occupational troops. Yoshino-san was one of the senior officers who knew Captain Yasuna Kozono well and could explain first-hand about what happened in this chaotic time just after the Emperor's formal radio announcement about Japan's surrender, which was broadcasted at noon, on 15 August 1945. Unfortunately, I never had the opportunity to meet this distinguished veteran in person, as he passed away before we could meet. Thankfully, through the assistance and kindness of the Yoshino family, I am able share his story with the help of his unpublished private diary and other documents. What occurred at Atsugi Air Base after 15 August 1945 and involved the 302* Kōkūtai, *led by Captain Kozono, may never be fully known. Nevertheless, I did my best to depict what happened at that time as told by the late Minoru Yoshino, members of the Kozono family, and any research materials that I could locate in Japan and the United States. Please note: given the amount of information and numerous sources utilized, this story will be presented in a slightly differently format than the others in this work.*

Yoshino's Background

Minoru Yoshino graduated from Sasebo Junior High School in March 1934. After joining the Imperial Japanese Navy, he attended the Naval Engineering College and graduated on 16 March 1938, as a member of the 46th Term Class. Throughout the Pacific War, the young mechanic served in both seaplane and flying boats units (like many of the men who would later join the 302 Ku*. Next, Yoshino-san took a training course on Kasei engines at the Yokosuka Naval Air Technical Arsenal. After this, on 8 April 1944, he became a member of the 302* Ku*, as a senior maintenance officer, and served with this group until the end of the war. His final rank in the Navy was maintenance lieutenant commander.*

Maintenance Lieutenant Minoru Yoshino took this formal picture just prior to getting married. (Minoru Yoshino Collection)

A casual photo of a recently married Yoshino-san when staying in Kamakura. (Minoru Yoshino Collection)

Captain Kozono & No Surrender at Atsugi

To better understand what happened at Atsugi Air Base after the 302 Ku *declared that they would continue fighting even after the Emperor's radio announcement, it is important to know more about their commander, Captain Yasuna Kozono. He was a traditional Japanese nationalist who believed that the Emperor was like a God and had to be protected at any cost. According to the 302* Ku *veterans I spoke with, Captain Kozono was totally trusted and loved by the men he served with. This twenty-five year Navy officer spent most of his adult-life in the service and served as an aviator on several aircraft carriers including* Kaga, Akagi, Ryūjō *and* Shōhō. *As for family-life, Kozono-san married his wife Haruko and they had four children together, two sons and two daughters.*

On 1 October 1941, he joined the famed Tainan Kōkūtai, *as the air officer, and on the 15th of that same month, Kozono-san was promoted to the rank of lieutenant commander. This unit saw extensive air combat beginning, on 8 December 1941, when they attacked the Philippines through the subsequent action in the Dutch East Indies and eastern New Guinea. On 1 November 1942, the Tainan* Ku *was renamed the 251* Ku, *and later that month, returned from the frontlines to regroup in Japan. Then, on 5 December of that year, he was promoted to be commanding officer of this renamed unit. The refreshed 251* Ku *returned to Rabaul, New Britain Island and were stationed at Lakunai Airfield (or east airfield). They participated in numerous aerial battles especially in the Central Province of Solomon Islands including the Russell and Rendova Islands. Commander Kozono served with the 251* Ku *until 25 September 1943. During the course of following year, he was as an instructor with the Yokosuka* Kōkūtai *and later a posting with the Yokosuka Naval Air Technical Arsenal. Lastly, on 1 March 1944, Captain Kozono became the commander of the*

302 Ku and did this duty until the end of the war. The Kozono family lived Kinugasa near Yokosuka while he was the commander of this unit. They lived in Kinugasa until moving back to Kagoshima in 1947.

Captain Kozono and his family during happier days. Pictured to the left of him is his wife Haruko. Three of their children are pictured here, left to right: their youngest daughter Sachiko, first son Haruo and elder daughter, Misako. Not pictured in this photo is their youngest son, Atsumi, who believes it was taken on 1 January 1944 in Gamagōri, Aichi Prefecture. (Atsumi Kozono Collection)

Right: Captain Kozono, sometime in March 1944, at Kisarazu Air Base. On 1 March 1944, the 302 Kōkūtai *was formed and this picture was taken there before the group moved to Atsugi Air Base on 30 March of that same year. (Atsumi Kozono Collection)*

Below: Vice-Admiral Michitarō Totsuka, the commander at Yokosuka, who got into a heated argument over the phone with Captain Kozono. This picture was taken just after he surrendered his base to Rear Admiral Robert B. Carney, U.S. Third Fleet Chief of Staff, on 30 August 1945. (NA)

According to Maintenance Lieutenant Commander Yoshino's memoirs, a few days before the Emperor's radio announcement on 15 August 1945, there were already rumors amongst the officers that Japan would accept the terms of the Potsdam Declaration and surrender. However, after the noon broadcast (on that same day) of the Emperor's decision to formally accept the terms of this allied declaration and cease combat operations, Captain Kozono simply could not believe that the Emperor really wanted to surrender – there had to be something more sinister going on. He felt that this wasn't true and instructed the men serving under his command to continue fighting. However, at that time, many sources state Captain Kozono was suffering from a second bout of malaria and very ill. Then, on the 16th or 17th of August, Yoshino-san happened to be in Kozono's office while running some errands when his commander had a very heated phone call with Vice-Admiral Michitarō Totsuka, (Commander-in-Chief, Yokosuka Naval District). In short, Captain Kozono yelled at the admiral and said that Emperor wanted to continue fighting; this must have been a trap by someone else. Kozono-san told Vice-Admiral Totsuka that he didn't have the "guts" to protect Japan and this is why the country was failing. Lastly, the fiery commander shouted to the admiral that his responsibility was to protect the Imperial Family. After hearing this conversation, Maintenance Lieutenant Commander Yoshino vowed to support Captain Kozono and fight to the end with him and his comrades in the 302 Ku.

Historian Shunsuke Sagara's book states that even one day before the Emperor's radio message (14 August 1945), Captain Kozono initially heard that the Japanese government would surrender so he sent a telegram to Vice-Admiral Kinpei Teraoka (Commander-in-Chief, 3rd Air Fleet) stating that he didn't agree with this decision. So, on the following day, the 15th, Captain Kozono declared to both his men and upper echelons of the Navy that they would continue fighting. This was viewed by the Navy and Japanese government as "Coup D'état" and that it must be suppressed. Next, on the morning of 16 August, Vice-Admiral Teraoka met with Kozono-san and tried to get him to accept the news of Japan's surrender. However, Captain Kozono refused and was sent to Yokosuka. In the meantime, the Navy assigned Captain Sakae Yamamoto to replace him. However, Lieutenant Yutaka Sugawara (Vice-Commander of 302) asked that Captain Yamamoto not come to Atsugi Air Base for a few days because this would only cause more troubles to an already chaotic situation. At the same time, members of the 302 Ku *were spreading leaflets over both public places such as bus stops and military installations to urge everyone to continue fighting.*

A tragedy also occurred during this same time period (15-16 August 1945), just after the surrender announcement. With Captain Kozono being incapacitated, one of his senior officers Lieutenant Kushichirō Yamada was in the position to handle the important affairs for the 302 Kōkūtai. *So when Prince Nobuhito Takamatsu called Atsugi to find out what was happening, Lieutenant Yamada was the person he spoke with. During this phone call, Yamada-san pretended that he couldn't hear the prince well because he was conflicted on what to do. Yamada-san wanted to follow Captain Kozono's orders but he knew the war was over. So, in the end, Lieutenant Yamada remained loyal to his beloved commander but felt strong guilt that he went against the wishes of the Imperial Family and Prince Takamatsu. On 21 August, the Lieutenant left the base and went home to his young bride Yuki. Together, Kushichirō Yamada and his wife committed suicide and their bodies were found together with the hope they would be together in their afterlives.*

What happened to Captain Kozono?

WHAT HAPPENED TO CAREER *Navy officer Captain Yasuna Kozono during the final days of World War II? As mentioned earlier, many sources depict him, on the 17 August, with having another bout with malaria and becoming very incoherent. Interesting timing for Kozono-san's health to take a turn for the worst? According to the late Haruo Kozono, eldest son of Captain Kozono, who I met at a 302* Ku *reunion in 2003, the malaria story is pure hogwash. In reality, his father was given laughing gas while he was asleep and then later morphine, which caused him to act so erratically. Haruo also said that he later learned that his dad was interviewed while in prison by Allied occupational forces. Kozono's late son stated that there were official postwar documents confirming this account of his father being drugged, but after a detailed search by this author, at the National Archives and other American-based institutions, nothing could be located.*

In the end, on 21 August 1945, while he was still incoherent, Captain Kozono was bound with both his hands and legs, and taken by car to Nobi Naval Hospital. Eights days after this, Kozono-san transferred to a naval prison at Yokosuka. On 16 October 1945, he was sentenced to life in prison, but on 5 December 1950, Kozono-san was released. He was completely pardoned on 28 April 1952 and moved back to Kagoshima that same year, but lost all his rights and honors as a former military veteran including his pension. Times were very difficult for the Kozono family. In October 1952, this once proud Navy officer was released from Kumamoto Prison (in Kyūshū) because of a special pardon as a result of the San Francisco Peace Treaty, which was signed by forty-eight countries and officially ended the war (even though it was essentially over in 1945). Three years later Kozono-san suffered a brain hemorrhage, which made writing difficult. He later recovered enough to begin writing his story of what happened at the end of the war, but would never finish this work because on 4 November 1960 at the age of fifty-seven, Yasuna Kozono suffered a second brain hemorrhage and passed away.

Regarding the restoration of Kozono's rights and pension, as a former officer in the Imperial Japanese Navy, this would require a lot of work in the postwar years, by his good friend Etsutaro Tanaka. In 1972, with the help of Shinji Kohama (who was a member of the Japanese House of Representatives); Tanaka-san brought this petition to that government body for review. Then, on 25 July, of that same year, Tanaka-san met with Prime Minister Kakuei Tanaka and asked for his advice on how to move forward and assist the Kozono family. Remarkably, Etsutaro Tanaka was able to get over 30,000 signatures from the public to help restore Kozono's rights and pension to his family and, in 1974, Haruko Kozono was finally able to receive some benefits (even though it was fourteen years after the death of her husband).

Meeting Prince Takamatsu

On 17 August 1945, while I was Atsugi Air Base, Captain Kozono had another bout with malaria; nearly unconsciousness with a high fever. Lieutenant Sugawara told me that the proposed enemy occupation of our base was very dangerous because the 5,000 men of our unit wanted to continue fighting the war. So, I was appointed to be a messenger and convey this message to Captain Chuzaburō Yamazumi (Chief of Staff, Third Air Fleet). Captain Yamazumi said that he understood how I felt but our headquarters already ordered to do whatever is necessary for Atsugi to be occupied by enemy troops and that these orders could not be changed. Surprisingly, I was ordered to stay on the base and personally welcome General MacArthur. In order to do this, I would be given two trucks and eight men who would help me. Captain Yamazumi said, "You should just stay on the base and make sure that American Army landed there. Once you see them land, you can leave Atsugi. The reason being is that if you are captured, you will probably be killed." I was grateful for him to tell me this because Yamazumi-san didn't want me to waste my life. This was a time of great confusion for both the Imperial Japanese Navy and our government so they really didn't know what to do. That is why Captain Yamazumi had to issue these orders to me. However, I was willing to die for Japan and when I thought of all these men who perished fighting for our country, I didn't want to "run away." Initially, this was an informal order but later, on 21 August, I was issued an official order to stay on our base as the stay-behind liaison officer.

As an anecdote, when I first heard the news that Japan might accept the Potsdam Declaration and surrender, I was a bit confused. I wanted to die fighting for my country and protecting the Imperial Family. Today, when some Japanese World War II veterans do interviews, they usually say that they were "relieved" when Japan surrendered. Personally, I didn't have those feelings. I felt "confused" that Japan was quitting the war and that I couldn't die for my nation. I just want people to know that not everyone was relieved that the war was over. Some of us were very shocked and wanted to continue fighting.

Prince Nobuhito Takamatsu, brother of Emperor Hirohito, while serving as a member of the Naval General Staff Office in Tokyo. He played an instrumental role in the peaceful handover of Atsugi Air Base to General Douglas MacArthur and his occupational forces. (Prange Collection, University of Maryland, via Michael Wenger)

On 19 August (before I was formally ordered to hand-over Atsugi Air Base), Prince Nobuhito Takamatsu, called Atsugi and said that he wanted to meet the people in charge. The following day, Lieutenant Sugawara and I went to the Department of the Navy, in Kasumigaseki, Tokyo, and met with the prince in an underground bunker. Our meeting lasted for about ninety minutes and he hoped to get our assistance for the formal surrender and occupation of our base to American troops. Prince Takamatsu spoke to us very calmly and didn't sound like he was under any duress. We were prepared to be scolded by him but the prince never did this. Instead, he spoke to us in a fashion that it appeared his position was actually lower than ours; it almost seemed like the prince was asked us for a favor. Therefore, I felt very obligated to follow his orders.

Prince Takamatsu went on to say that his orders were based on the Emperor's "100 Year Plan for Japan." He said that the Emperor told him that right now is the time to be "patient" and follow the rules by the American occupation troops. The Emperor also said that even if you have to shine their shoes, you should obey them. He also firmly believed that Japan would one day have a strong economy and be a country of peace that was rich in culture. When I see Japan today, it has become an economic giant. Remarkably, one-hundred years have not even passed since this happened. So, I am very impressed that our Emperor this foresight (at the end of the war). Keep in mind when I heard all of this from Prince Takamatsu, there wasn't any food in Japan and most of our country was literally burned to the ground because of the war. At that time, you couldn't imagine a future that Japan has today; this really impresses me.

The prince also emphasized that life was precious and he didn't want any more Japanese people to waste their lives needlessly. He also said, "Maybe, you think that it is better to die now since Japan surrendered. Instead, I want you to live and help rebuild our nation." Even though we didn't say anything to the prince to indicate our true feelings (about wanting to die for our country), he knew what we must have been thinking. Prince Takamatsu went on to say that he wanted to see both of us again after everything settled down and Japan was stable again. The prince's postwar books were very critical of the Japanese military and his underlying theme was that so many precious lives were wasted for nothing because of the war. In 1959, I met him again and he was very happy to see me. After this, I became a lot closer to our Imperial Family and heard lots of stories from them. I feel Japan is very fortunate to have such a good family helping rule our nation. For as long as I live, I feel that protecting the Imperial Family is very important to do. It is regretful that Captain Kozono wasn't well enough to meet either the Emperor or Prince Takamatsu because he strongly believed that surrendering wasn't the Emperor's idea. If he was able to meet either of them, the "Atsugi incident" would have never occurred.

*****

After meeting Prince Takamatsu, Maintenance Lieutenant Commander Yoshino and Lieutenant Sugawara returned via a car to Atsugi Air Base. Following dinner, that night, 20 August, all of the officers from the 302 Ku *got together. Lieutenant Sugawara said, "We feel bad about Captain Kozono but it was the Emperor's decision to surrender so we must convince our men. Yoshino-san and I accepted this order from Prince Takamatsu so if any one of you are not with us, please kill us. Otherwise, we would not be following orders." None of the officers spoke and then everyone left the room. According to Maintenance Lieutenant Commander Yoshino, he and Lieutenant Sugawara were expecting the worst and the night was spooky. Perhaps, some of the young soldiers in the unit would try to kill them because they wanted to continue fighting. In the end, nobody came to see them, so both Yoshino-san and Sugawara-san returned to their quarters for the night.*[1]

At 0800, the following morning, on 21 August, all of the men in the unit got together and Lieutenant Sugawara gave the order to begin disarming all of the aircraft. This command included getting rid of all the ammunition and fuel so that there would be no more combat missions for the 302 Ku. *So, Yoshino-san went back to his quarters and starting cleaning out his military documents, etc. Just as he was doing this, somebody knocked on his door and said that there was an emergency because lots of the unit's aircraft began to take off. Thirty-three aircraft departed Atsugi and flew to Kodama and Sayama Air Bases (both in Saitama Prefecture). The overall leader of this group was Lieutenant (junior grade) Ryōji Iwato.*[2] *Many of these aviators were members of a group called* Kikusui-Tai *and hoped to join liked-minded Army veterans, who wanted to continue fighting. Out of the group of eighteen aircraft that were bound for Kodama Air Base, one* Zero *fighter flown by Lieutenant Yoshinori Kaida never went there. Instead, this lieutenant flew over to his parents' home, did a low pass and waggled his aircrafts wings to say goodbye. Then, Lieutenant Kaida went to the Tokyo Imperial Palace (home of the Emperor), circled over it a few times, and after this, he crashed his* Zero *into Tokyo Bay. The rest of the seventeen planes landed at Kodama Air Base. The other portion of the original group of thirty-three aircraft consisted of fifteen planes, which all landed at Sayama Air Base. As an interesting note, no* Raiden *fighters took part in this rebellion because they were no longer flyable at Atsugi Air Base since their batteries, fuel and props already had been removed. Remarkably, one of mechanics who did this decommissioning work on the* Raidens *was Lieutenant Yoshimasa Yamamoto, the first son of Admiral Isoroku Yamamoto.*

Around this time, another group of 302 Ku *men, between thirty-fifty mechanics and others, traveled to Sayama Air Base via ground transportation and joined the aviators that flew there earlier. However, by this time at both Army bases, the troops stationed there were ready to give up the fighting so nothing materialized after this. In desperation, Lieutenant (junior grade) Iwato flew to Kodama and went to see Major Motomichi Uki, army commander of the 98th* Hikō Sentai, *which was based at this airfield. Major Uki told him that his men could not be involved in this planned revolt unless there were other movements like this (to continue fighting) around Japan. Iwato-san agreed to Uki's plan and, in the meantime, the group of the 302* Ku *men was able to stay at a local high school. Local farmers helped provide food and sleeping provisions for Iwato-san and his comrades.*

On 22 August, Vice-Admiral Teraoka flew to Atsugi Air Base and was able to get a detailed report of what was happening from the newly assigned 302 Ku *commander, Captain Sakae Yamamoto. Initially, Teraoka-san sent Captain Chihaya Takahashi, (3rd Air Fleet, Chief of Staff), to Sayama Air Field to try to persuade the 302* Ku *men to stop, but he was unsuccessful. Next, the admiral dispatched 302* Ku *Vice-Commander Lieutenant Sugawara and fellow* Raiden *pilot Lieutenant Sumio Teramura, who brought the message if the men returned to Atsugi, they wouldn't be prosecuted.*[3] *After these fellow 302* Ku *officers visited, some of the aviators participating in this last-ditch attempt to continue fighting changed their minds. On that same day, around noon, fifteen* Zero *fighters and one* Suisei *returned to Atsugi Air Base while an additional* Ginga *went to Fujieda Navy Air Field to give up the fight. Nevertheless, there will still some others who stayed at Sayama along with the other group at Kodama Air Base.*

Vice-Admiral Teraoka sent Lieutenant Yasuo Shimizu (a 302 Ku Hikōtaichō*) to Kodama to talk to the remaining 302* Ku *holdouts, but initially, Shimizu-san couldn't get there because of a typhoon. It was during this typhoon that Vice-Admiral Teraoka formally demobilized the once proud 302* Kōkūtai *at Atsugi Air Base and some of the unit members shot their weapons in fury before leaving the base. Next, on the following day, Lieutenant Shimizu, with the support of about ten men, flew on a Type 1 land attack plane to Kodama. From the air, he noticed about twelve 302* Ku *aircraft on the field with only a handful of men nearby. So, Shimizu-san quickly landed and ordered all the tires to be cut on these rebel aircraft. Lieutenant (junior grade) Iwato ran from the school where he was staying but it was too late to do anything; his airplanes were no longer airworthy.*

Back at Atsugi Air Base, Vice-Admiral Teraoka was really concerned what Lieutenant (junior grade) Iwato and the others may do next; maybe they would try to occupy central Tokyo or attack some of the American ships in Tokyo Bay? With concerns such as this he sent Lieutenant Teramura, Captain Chihaya Takahashi and Lieutenant Commander Seijirō Takagi back to Kodama Air Base to try to talk with Iwato-san and the others once again. Finally, Lieutenant (junior grade) Iwato and the men involved in this Coup D'état agreed to stop after Captain Takahashi promised them that they could talk with the Emperor directly about their concerns. After this, Lieutenant Teramura went to Sayama Air Base and they agreed to stop fighting; some of these men returned to Atsugi but a few just disappeared. The following day, on 25 August, four buses came to pick Iwato-san and his group up, but instead of going to the Tokyo Imperial Palace, they were put into custody by the military police. At 1500 that same day, 302 Ku's *revolt was over.*

Out of all the 302 Kōkūtai *personnel who were involved in this incident just sixty-nine men remained because many others had given up the fight by this point. Forty-four of the petty officers who were involved with this incident were sentenced with a one year prison sentence and three years of probation. As for the twenty-five officers involved, their jail sentences were for four to eight years, but they were later released with a special pardon on 3 November 1946, when Japan passed a new constitution.*

Handing over Atsugi Air Base to the United States Army

On 22 August, the 302 *Ku* demobilized and all of the higher-ranking unit commanders were ordered to Fujieda Air Base (in Shizuoka Prefecture) and stay there until they received further orders. As for me and the eight other men, all of us were suppose to stay at Atsugi Air Base until the American troops landed there. Personally, I had several responsibilities besides staying on Atsugi Air Base and handing it over the occupational forces. These duties included arranging means of communication with the Americans and doing an inventory of everything on the base. It was assumed that I would be captured and killed doing these duties so this was a very dignified thing to do.

The United States Army troops started to arrive at Atsugi Air Base on the 28th of that same month. There were countless aircraft and vehicles coming in. All of the aircraft had their guns pointed us; this gesture gave us the message that they would shoot anyone who tried to resist. Since they were flying at such low altitude I could clearly see the faces of the men; this made me very angry. I realized that the enemy would occupy our land and that we had lost the war. Why did we fight so long just to lose? I thought about all of the people who had already died in combat. What were they thinking when seeing all of this? Did they die for nothing? But, at the same time, I remembered what Prince Takamatsu had told me and the promises that I had made to him. It was my personal duty to allow the enemy troops to occupy our base and that I was following these orders from the Imperial Family. So, I actually felt "relief" on the day that I was able to welcome the occupation troops on our base.

*****

The first American aircraft, C-46 transport, touched down at 0828, on 28 August, at Atsugi Air Base. It was followed by fifteen others, which were transporting the initial advance occupation forces of just under 200 men, who were led by Colonel Charles P. Tench, GSC, of the G-3 Section of GHQ. Colonel Tench was met by Captain Chuzaburō Yamazumi, Lieutenant Generals Seizo Arisue and Senichi Kamada along with Ken Tsurumi of the Foreign Office. This initial group of American troops was sent to prepare Atsugi Air Base for the larger-scale occupational forces that were expected to arrive in the coming days.[4]

Yoshino-san in his typical "work uniform" while serving with the 302 Ku. *(Minoru Yoshino Collection)*

30 August 1945, General Douglas MacArthur disembarks his aircraft at Atsugi Air Base. While serving in Japan, the general has the title, The Supreme Commander for the Allied Powers (SCAP) and overseas the occupation of the country in the initial postwar years. (United States Army)

The Arrival of General MacArthur at Atsugi Air Base

Two days later, on 30 August 1945, the much larger elements of the 11th Airborne Division flew from Okinawa and began landing at Atsugi Air Base in the early morning hours. Aircraft landed every three minutes throughout the day and were loaded with both troops and cargo that were needed for the occupation. After Major General Joseph M. Swing, (Commander of the Division), arrived, he began planning with Lieutenant General Arisue, for the expected later arrivals at Atsugi, General Robert Eichelberger, (Commander of the Eight United States Army), and lastly, General Douglas MacArthur. Just before 1400 on that same day, General MacArthur landed at Atsugi and was warmly welcomed by Generals Swing and Eichelberger along with the men of the 11th Airborne Division. MacArthur proclaimed "Melbourne to Tokyo was a long road but this looks like the payoff," before boarding a waiting car to take him to Yokohama, where a temporary General Headquarters was being set up.[5]

*****

On 30 August, there was supposed to be nine men (including me) to hand over Atsugi Air Base. But, I was there with only one other guy. I assume the seven other men ran away because they didn't want to die. Even the newspaper reported that the two of us were all that was left of the 302 *Ku*. I had prepared my sword and a cyanide capsule with the assumption I would die on this day. This was the day that General MacArthur and other important dignitaries arrived at Atsugi. The proud expression on MacArthur's face said, "Look who is talking now?" Since there was a two-day cushion period between when the initial enemy troops arrived before General MacArthur came, I got used to seeing Americans being there. So, I was able to welcome him in a calm manner. Even though, I was still worried that MacArthur would have control over the Emperor and our government, I still did feel some relief that I could fulfill the promise made to Prince Takamatsu.

Reflections

In March 1945, large formations of B-29 bombers were bombing many cities in Japan so twenty 302 *Ku* aircraft were sent to Kanoya Air Base to try to counter these enemy aircraft. None of our *Raiden* fighters were shot down but I recall seeing these young Navy *tokkō* (not 302 *Ku*) pilots that we sent off on their suicide attack missions. All of these aviators were in their 20s or younger and wore *hachimakis* on their heads which said, "Victory!" They looked very strong and had a powerful presence. Before this, I had already experienced seeing a lot of men die from our unit, but this was different. These young men flew the *Ohka* which was never meant to return. In my mind, I thought, "We will be right behind you. Let's meet again at the Yasukuni Shrine."

Even today, I can't forget the feelings I had so many years ago. I still dream about these young aviators as well. We need to remember to praise and respect them for the sacrifices they made. So, what about modern-day Japan? These days, dying for your nation is considered taboo. People only think about themselves and not others; this troubles me very much. What would the Japanese people who died fighting in World War II think of this?

[1] 1 August 1959, *Maru* magazine.

[2] He was a member of the 73rd Class, Naval Academy.

[3] Lieutenant Teramura was a member of the 71st Class, Naval Academy.

[4] www.history.army.mil

[5] GHQ SWPA & USAFPAC, Mo Sum of Opns (S), Aug 45, p. 12.

Chapter 17

"I can't forget the sense of loss that was left in our minds after the deaths of our comrades."

Lieutenant Ryōichi Yamada (IJN)

I recall very vividly my first meeting with former World War II Imperial Japanese Navy fighter pilot Ryōichi Yamada. He was very intelligent, and spoke in a "matter of fact" way when talking about his wartime experiences along his postwar military experience (until Yamada-san retired as a general). I believe that many of the veterans which I met who saw a lot of combat share the latter attribute while sharing their experiences with me. I feel very fortunate to have met this amazing gentlemen and could hear his stories first-hand; truly amazing. Special thanks for historian Henry Sakaida's assistance with this story.

*****

Yamada-san Joins the Navy

I was born on 3 September 1923, which was two days after the great Tokyo earthquake. On 1 December 1939, I joined the Naval Academy, as a member of the 71st Class. While attending the academy, all of the cadets were tested for proficiency in aviation and the tests determined that I would be an excellent candidate to become a pilot. I really liked the idea because I did not want to get seasick living on a boat all of the time. I also thought that flying would be a suitable match for my personal skill-set as a good all-around athlete, *sumō* champion and sprinter. Finally, the last thing that made me an excellent choice for being an aviator was my eyesight, which was 3.0. Due to my incredible vision, I was able to see aircraft, which were ten miles away while most people could only see a distance of about five miles.

At the Naval Academy, the day would begin at 0530, with about one hour of exercise, which could include activities such as fencing, judo and boating. At 0800, classes would start and, after lunch, there would be additional classes in the afternoon. There also would be more exercises such as swimming, boating, *sumō* wrestling and gymnastics.

Normally, the curriculum in the academy took four years to complete, but I finished in three. During the second year, cadets were required to learn a foreign language, math and physics; I chose English. After completing the second year, the classes became strictly military oriented. For example, a course that would have been taken at this time was gunnery. There was one important difference between the Army and Navy academies. In the Army Academy, the classes were mostly about military history and tactics but the Naval Academy had no such classes. This was done on purpose because the Naval Academy didn't want their cadets to be caught up with military theories. After the war, I met former Army Academy graduates and spoke to them about their training. Then, I changed my mind after talking with the Army veterans, and I now feel that the Naval Academy should have taught similar classes in their curriculum.

After graduating from the Naval Academy on 14 November 1942, until 1 June 1943, my classmates and I where placed aboard fleet vessels for training. On 1 June 1943, we were ordered to return, but it took one month for the ships to get home from the various assignments. Then, in July, my flight training commenced. The criteria for being chosen for a flying position was simple. While you were in the academy, tests would be given to see what position you would be best suited for, based on your skills. You were also allowed to submit your request; nevertheless, the Ministry of Navy had the final say over the decision.

From 1 July 1943, until mid-February 1944, I trained on the Type 93 Intermediate Trainer. After that time until the end of June, I practiced flying in *Zero* fighters. My total flying time on the Type 93 was around 150 hours while on the latter aircraft was fifty-five hours. Typically, we had to train an additional fifteen hours, but because of the dire war-situation, the training was cut short. *Buntaichōs* from various air groups would come over to the training school and ask for new pilots. For example, during one of my gunnery training sessions, the aviator trainees were separated into two groups, but only one of them actually received the training. Therefore, my group never fired any live bullets.

Off to War

On 1 July 1944, the young pilot was assigned to the Fighter 402 Hikōtai, *which was a division of 341* Ku (Shishi *or "Lion"). During this time the group was led by Captain Motoharu Okamura, but in October, Commander Tadao Funaki took over. They flew the new, temperamental Kawanishi* Shiden *fighter. Nine days later, on the 10th, this air group was reorganized into two divisions, the Fighter 402 and 401* hikōtais. *Lieutenant Iyozō Fujita was the commanding officer of the Fighter 402* Hikōtai *and Lieutenant Ayao Shirane led the Fighter 401* Hikōtai. *The last compliment of the 341* Ku *was Fighter 701* Hikōtai, *initially led by Lieutenant Commander Hideki Shingō, but in August, Lieutenant Kunio Iwashita replaced him. This respective* hikōtai *was utilized as the T-Attack Force of the unit and consisted of several different types of aircraft.*

A wartime picture of fighter pilot Ryōichi Yamada in his flight gear. (Henry Sakaida Collection)

The main group of Fighter 401 Hikōtai *transferred to the city of Kaohsiung, on the island of Formosa during late August and September of 1944. The squadron's mission was to fight the allied aircraft flying from China. Unfortunately, only a small number of the squadron's* Shiden *fighters were operational, and difficulties were encountered during training. Towards the end of September, Fighter 402* Hikōtai *transferred to Miyazaki Air Base, on the island of Kyūshū.*[1]

*****

UNFORTUNATELY, IN LATE 1944, I had a serious landing accident during a ferry flight of a *Shiden* fighter. Frequently, the brakes on the aircraft would not work, then all of sudden; they would catch and lock up. This problem caused the *Shiden* to do a "nose-over." I survived this accident because I held on to the bottom of the instrument panel when the aircraft flipped over. Prior to this most pilots did not survive this type of crash. After learning how I survived the accident, Japanese authorities decided to make this an emergency procedure for this type of crash. After implementing this emergency procedure, miraculously nobody else was killed during a "nose over" crash in a *Shiden* fighter. I had a twenty-five centimeter wound to my head so even when I smiled, my expression would look funny because my skin was stretched.

Yamada-san bandaged up after his landing accident in late 1944. (Ryōichi Yamada Collection)

*****

WHILE YAMADA-SAN WAS still hospitalized, the 341 Ku *flew combat missions over Formosa and the air group had some shoot-down claims. Unfortunately, according to this veteran, the 341* Ku *didn't have much success on these missions. In late October, the air group then moved to the Philippines and consisted of thirty-six fighters, but only twenty-one aircraft were flyable. Unfortunately, the 341* Ku *suffered heavy losses on their 24 October mission and was reduced to only four flyable fighters. Furthermore, many of the unit's pilots were inexperienced, so many of the men had trouble flying the new, temperamental* Shiden *fighter against allied aircraft such as the P-38 and F6F Hellcat.*

*****

AFTER I RECOVERED FROM my crash injuries, I flew via a transport plane from Miyazaki to Naha (Okinawa). On Okinawa, I found two *Shiden* fighters that were left there because of engine problems. I flight-tested them and found that the aircraft were fine. Around 28 October 1944, I flew from Okinawa to the Philippines via Formosa. Our air base was named Mabalacat, which is one of the southernmost airfields, located on the island of Luzon. It was used by both the Army and Navy, but the Army ran the housing facilities. The living conditions on this base were deplorable. Our living quarters were shabby huts, which were built with a high floor. Typically, two officers shared an eight *jō* (a 142 square foot) room. The largest quarters were about twenty *jōs* (356 square feet). The *Shiden* fighter only operated from this air base. Unfortunately, many of the men were inexperienced, so this aircraft could not be used to its full potential. Furthermore, due to a lack of spare parts, the majority of the *Shidens* were not flyable.

On 12 November, a group of *Shiden* fighters participated in a mission. I was the first person scheduled to take off, but the aircraft would not start. I tried the hand pumps, but it still would not start. I then noticed that the fuel cock was off, so my fighter was the last one to take off. First, I thought that the Grumman fighters were getting shot down, but it was actually the Japanese! I decided not to fly to higher altitude because my speed would be lost. Instead, I escaped the battle scene by flying low; otherwise, my aircraft would have been shot down.

During this time period in the war, due to bad communication equipment, lack of radar, radios and watch stations, we were always had enemy surprise attacks. I recall seeing F6Fs, P-38s, B-24s and B-17s. We were always taking off after being attacked. The enemy also was increasing in numbers. As I mentioned earlier, the *Shiden* fighters were stationed at Mabalacat Air Base, but there also were various *Zero*-equipped air groups in the Philippines. Furthermore, additional air groups which flew the latter fighter were located at the central field in Clark, the north field in Clark and Mabalacat Air Base (further north of Clark).

343 *Kōkūtai* Days

Upon his return to Japan, Yamada transferred into the newly formed 343 Kōkūtai, *which was named* Tsurugi *or "Lance." It was formed on 25 December 1944. The unit was comprised of five squadrons including the 401, 301, 701, 407 (fighter* hikōtais*) and the 4th Reconnaissance* Hikōtai. *Captain Minoru Genda, the man behind the infamous Pearl Harbor attack, led the unit. He was a pioneer in Japanese aviation and, even before the infamous Pearl Harbor attack, promoted the use of the carrier instead of the battleship for naval combat. Genda-san believed that the only way to stop the allied air armadas was to use the best fighter available, which was the Kawanishi* Shiden Kai, *and have only experienced pilots fly in his unit. The backbone of the squadron consisted of several aces including Lieutenant Naoshi Kanno, Ensign Kaneyoshi Muto and Chief Petty Officer Shoichi Sugita. He also believed in good warning systems so Captain Genda relied upon his 4th Reconnaissance* Hikōtai, *which would gather information using a high-speed carrier observation plane called the* Saiun *("Iridescent Cloud"). He always believed in the improvement of air-to-air communications and formation combat.*

Remarkably, according to Yamada-san, the 343 Ku *was the only unit to have properly working radios in their aircraft in the Imperial Japanese Navy. He said that the local Yokosuka* Ku *discovered that simply adding resistors to the spark plug wires made the Type 3 radio in their* Shiden Kai fighters *work properly! Yamada-san went on to say that soon after hearing this, Genda-san ordered his entire squadron of aircraft to be fitted with the resistors to improve air-to-air communications, and within a day each plane was outfitted with them.*

Legendary naval aviator and tactician Captain Minoru Genda (left) was the commander of the 343 Kōkūtai *and his executive officer was another experienced airman, Lieutenant Commander Yoshio Shiga. (S. Bon Group)*

Picture taken on 10 April 1945 at Matsuyama Air Base just before Fighter 301 Hikōtai *of the 343* Ku *departed for its new base at Kanoya. In the foreground is* Shiden Kai *fighter (343-A-15) which was often flown by Lieutenant Naoshi Kanno. (S. Bon Group)*

Captain Genda utilized some simple expressions to boost the morale of his men. He gave each of his squadrons popular names. The Fighter 301 Hikōtai *was called the* Shinsen-gumi *"Elite Guard," the Fighter 407* Hikōtai Tenchugumi *"Heavenly Punishment Unit," the Fighter 701* Hikōtai Ishin tai *"Imperial Restoration Unit," and finally, the 4th Reconnaissance* Hikōtai, *the* Kiheitai *"Commando Unit." However, it is not known if the Fighter 401* Hikōtai, *which was the training squadron, had a name. The Fighter 401* Hikōtai *consisted of a large number of late term flight school pilots. The Fighter 301* Hikōtai *was led by veteran pilot Lieutenant Naoshi Kanno, and suffered the most casualties out of all five of the squadrons. According to well-known Japanese ace Ensign Saburō Sakai, Kanno was much too reckless in combat and a bad leader, so his squadron had the most losses. Lieutenant Takashi Oshibuchi led the Fighter 701* Hikōtai, *and it suffered the least amount of losses. Yamada-san was a member of this* hikōtai *Finally, Captain Genda posted banners in front of the 343* Ku *barracks to provide inspiration to his men.*

On 18 April 1945, the 343 Ku *transferred to Kanoya Air Base after the allied invasion of Okinawa. The unit participated in* Kikusui *"Floating Chrysanthemum" Operation Number Two, but attacks on the island of Okinawa were difficult due to the limited range of their aircraft. They also helped protect the* tokkō-tai *pilots on their missions to Okinawa. Between 18 April –11 May, the 343* Ku *was based at Kokubu Air Base No.1, because of the constant B-29 air attacks over Kyūshū.*[2]

*****

I DIDN'T FLY ANY *tokkō* escort missions with the *Shiden*; instead we flew the *Shiden Kai*. They were escort missions, but we would not directly cover them. There was a range problem. Our aircraft could fly to Amami Oshima, and we could fight the enemy there. Our fighters did not have the range for direct cover. Unless you were already at high altitudes, it was very difficult to catch the B-29s. Furthermore, the high altitudes made fighting the B-29s very difficult. The *Raiden* fighter was a better plane to fight the B-29s, because it had better climbing capabilities. We did have radar though. If the early warning system worked, we would be able to get up there before the bombers came, although the chance of success was slim. When we were chasing after the B-29s, it was impossible to shoot them down.

Combat over Kure

On 3 May, the 343 Ku *then transferred to Ōmura Air Base. They intercepted allied aircraft in the Goto Islands region. Unfortunately, the unit suffered heavy losses of supplies due to the constant Allied air attacks, which increased after the capture of Okinawa. Furthermore, they also lost a number of experienced pilots due to combat losses. It was the beginning of the end for the unit and Japan itself. For Yamada-san, one of the most eventful missions occurred during the allied attack of the great Japanese naval city of Kure. The American Task Group 38.1 was assigned to attack the city during July 1945. During the war, Kure had a population of over 400,000 people and was the greatest Asian military port at that time. It also was a premier ship building center for the Japanese, and the* Yamato, *the country's ultimate battleship was built there.*

*****

On the morning of 24 July 1945, when the tide of war was unfavorable, about 500 carrier planes from the enemy task force approached Tosa heading northward and were in striking distance to the Kure naval harbor area. Upon receipt of the information, the 343 *Ku* decided to intercept them with all of our airworthy *Shiden Kais*. However, we only had twenty-one fighters because of heavy fighting and insufficient replenishment. In spite of such unfavorable conditions, Commander Genda still said this, 'The 343 *Ku* should stop the enemy offensive by destroying their planes one after another, if we find them.' At 0904, twenty-one *Shiden Kai* fighters, from all three squadrons took off from Ōmura Air Base with the leader of the Fighter 701 *Hikōtai* leader Lieutenant Takashi Oshibuchi, in the lead. Lieutenant Oshibuchi was a member of the 1st student class (68th term) when I entered the Naval Academy. Our relationship was not a simple one like that of a senior and junior or a squadron leader and a division leader but a close relationship between both a 1st class and a 4th class student.

We trusted each other. He was a unique fighter pilot in the Imperial Japanese Navy in the sense that there was no other officer who was so trusted by both his seniors, and men. It was largely due to his splendid leadership. All of the pilots in his squadron were determined to "die together" with our leader. Lieutenant

An aerial reconnaissance photo of Kure naval harbor area filled with Japanese vessels. (Henry Sakaida Collection)

Lieutenant Takashi Oshibuchi during his 343 Ku *days. (Henry Sakaida Collection)*

Oshibuchi was a warrior of gentle character. However, once in a fight, he demonstrated his persistent fighting spirit. Oshibuchi-san went to the forward position during any combat engagement. His skill for leading a large formation in the air was superb. He was very popular among his men, and they were proud of him. Often, his men were boastful of him to other members of the 343 *Ku*. Because of the atmosphere that he created around us, we strangely had no fear of 500 enemy planes attacking our squadron in combat.

Under the command of Lieutenant Oshibuchi, the twenty-one *Shiden Kais* waited for the enemy planes over the Bungo Strait. Suddenly, the voice of Commander Genda came on the wireless phone giving us instructions. Soon numerous spots appeared before my eyes, and they looked like scattered sesame seeds. Luckily, the enemy did not see us, and they kept heading south towards their carriers. After finishing their raids on Kure, formations of about thirty planes each were coming towards us in short intervals. There was a reported number of 500 aircraft, so there would be flights of ten, and several additional echelons. Suddenly my eyes were filled with half of them, probably two hundred and fifty aircraft! It was an overwhelming sight!

We did not intend to attack the enemy aircraft from the front because, even with the strength of twenty-one *Shiden Kais*, we would be overwhelmed by the enveloping attack from succeeding flights. Our squadron decided to attack the rear of the formation. It would take some time for the proceeding formations to turn back to aid the aircraft under attack. In the meantime, we would destroy the target formation and retreat like a bolt of lightning! We wanted to attain the maximum effect with the least cost. However, it was difficult to implement our tactics because of the limited flying time of our *Shiden Kais*. Hence, Lieutenant Oshibuchi decided to attack a formation that was flying at a longer interval between it and the proceeding formations.

We made a steady and careful approach, trying not to be noticed. Our pilots took a position almost straight ahead of the enemy formation, when Lieutenant Oshibuchi suddenly nosed down. It was a diving attack from behind and above at an altitude of 6,000 meters (19,684 feet). Simultaneously, after seeing Lieutenant Oshibuchi begin to dive, his signal to attack, the planes of both the Fighter 701 and Fighter 407 *hikōtais* charged and we were brought into a swirl of heated fighting. Flying in line with the lead plane, I also fired my first burst of the twenty-millimeter cannons at an enemy plane and then climbed to prepare for a second burst. At the same time, the enemy formation became disordered, and the *Shiden Kais* flying cover began to attack the succeeding enemy formation that approached. The battle became a large melee in an instant!

During the fight I was separated from Oshibuchi's plane. I glanced around for him during the melee and was relieved to see Flight Chief Petty Officer Jirō Hatsushima, the third man of our flight, escorting his plane tightly. I thought 'He would be all right if Hatsushima is covering him,' because he was an excellent pilot. During all of the confusion with planes crossing back and forth, I caught a glimpse of our leader's plane, together with Hatsushima, dog fighting at quite distance from me. That would be the last time that I saw his plane. At the same time, I was busy gripping the shooting knob of my aircraft and thinking, 'Be careful not too far' while battling the enemy aircraft.

A while later I stopped fighting because of a lack of fuel and ammunition, so I returned to Ōmura Air Base. I waited a long time, but Lieutenant Oshibuchi and CPO Hatsushima did not return. We destroyed sixteen enemy planes, which didn't include another two or three planes that were added after doing research later. Unfortunately, six pilots were lost in return, including Lieutenant Oshibuchi and CPO Hatsushima. On the same day, the 343 *Ku* received an Emperor's commendation for service. However, I can't forget the sense of loss that was left in our minds after the deaths of our comrades. So, celebrating with *saké* was not the right thing to do that day.

*****

The official American losses indicate twenty aircraft for their raid on Kure, but the actual losses were twenty-six aircraft. The 343 Ku *lost six of their own men, but they put in claims for sixteen U.S. planes destroyed. According to the original 343* Ku *records, Oshibuchi made three firing passes on the enemy aircraft and was hit in the engine. He then dived down, became separated during the heavy fighting, and was lost. Hatsushima was simply listed as MIA. Besides the terrible loss of Oshibuchi and other pilots, the unit also lost ace Ensign Kaneyoshi Muto. Muto, nicknamed* "Kin-Chan" *which means* "Little Gold," *participated in heavy fighting over the Solomons, Philippines, New Guinea, China and later home defense over Japan. In the 343* Ku, *he was the leader of the 301* Hikōtai, *and his death was a great lost to the unit.*

American participants from VF-49 also have vivid memories of their mission on 24 July 1945. Please note: both of these participants misidentified their opponents as flying Franks, *which was the allied nickname for the Japanese Imperial Army fighter, a Nakajima Ki-84, but in actuality they were engaging Navy Kawanishi* Shiden Kais, *Allied code name* "George" *aircraft.*

Ensign Kaneyoshi Muto was one of six 343 Ku *losses on 24 July 1945. (Henry Sakaida Collection)*

Lieutenant (junior grade) Jack Gibson recalled this air battle – "We heard a call from some fighters ahead that they were jumped by enemy planes. We immediately joined up to go to their aid. As we approached the planes ahead, I noticed two planes coming toward us, and I went with my wingman, Lieutenant (junior grade) Williams, to investigate. Their planes had wing tanks, and as we approached, they dropped them and started to dive for the water. By now we could see that they were '*Franks*,' and we started to close on them. Our section split, and I followed one while Lieutenant (junior grade) Williams closed on the other. The F6F-5 had no trouble at all in closing on the *Frank*, with the addition of water injection. The Jap tried to turn inside me, but every time that he started a turn, I would fire a short burst in front of him, which would always bring him back straight and level. When I got in effective range, my first burst knocked his wheels down. I overran him and had to pull up in a wingover to drop back on his tail. He then tried skidding turns without success. By this time, I began to get my bursts into the cockpit, and he commenced to smoke. He finally fell over on one wing and went straight into the water without burning."

Lieutenant (junior grade) George Williams had this to say about the melee – "Lieutenant (junior grade) Gibson 'Tally Ho'd' two enemy fighters under us on the water. We made a 180-degree turn and gave chase, each taking one. I closed rapidly on my plane, which I am reasonably sure was a '*Frank*.' Just as I came into range on his tail, he executed a very tight turn, which I attempted to follow, but I was not successful. Before I blacked-out, I shot about a three-second burst, and I think that it killed or wounded the pilot. When I came to, I found that he was falling in a tight spiral trailing a great cloud of smoke. Just before he struck the water, another group of friendly fighters followed him down and fired into his plane."

Above: A wartime picture of VF-49 pilot Lieutenant (junior grade) Jack Gibson (front row, far left). (Henry Sakaida Collection)

Right: VF-49 aviator Lieutenant (junior grade) Williams. (Henry Sakaida Collection)

*****

After their fateful mission of 24 July 1945, the 343 Ku *did not participate in any further large-scale aerial battles, mostly due to a lack of experienced pilots. With heavy losses to the 343* Ku*, including the veteran pilots Oshibuchi, Kanno, Muto and Sugita, fighting the vast allied air armadas became increasingly difficult. Nevertheless, the unit earned its place in the history books. According to Genda, the unit accounted for about 170 enemy aircraft shot down, with a loss of seventy-four pilots to the 343* Ku*. This was all done in a six-month period, with limited supplies, while flying against attackers that were always in superior numbers.*

Postwar

On 1 April 1949, Yamada-san married Oshibuchi's sister. He also returned to his love, flying, and had a distinguished military career in postwar Japan. On 1 July 1954, Ryōichi Yamada joined the Japanese Air Self-Defense Force. He trained to fly on American T-33 jets and later went to the United States to participate in a military aid program for foreign countries. During the program, he flew both F-86 and T-33 jets. In May 1956, Yamada-san returned to the

United States and went to Nellis Fighter Weapons School. Upon returning to Japan, he became a flight instructor at Hamamatsu Air Base. In 1959, the longtime aviator became Unit Commander at Chitose Air Base and then went to Air Staff College for ten months. After completing his courses at the college, he returned again to Hamamatsu and became the flight-training leader. He then became an instructor at the Japanese War College, and later served in various commanding posts. Ryōichi Yamada was part of the selection committee for the next generation of fighters after the F-104, which was the F-4 Phantom. He was probably the only person in the postwar Air Self Defense Force to serve as a group chief, section chief and department chief in the course of his duty. After a lifetime of service to his country, Yamada retired from military service in 1981, as Chief of Air Staff Office (general), the highest rank in the Self Defense Force. This amazing veteran passed away on 27 February 2013.

The author with veteran Ryōichi Yamada after an interview in Tokyo. (Ron Werneth)

[1] The background history of the 341 *Ku* was derived from the excellent *Japanese Naval Aces and Fighter Units of WW II.*

[2] Likewise, the background history of the 343 *Ku* was taken from *ibid*, 198-202 and with the help of historian Henry Sakaida.

Chapter 18

"I began to feel that the only way to fight back against such a tenacious enemy was to become a pilot"

Lieutenant Morimasa Yunokawa (IJN)

After I got "serious" about learning more about the Japanese side of the war, I started reading about their attempts to change the tide of the conflict by doing tokkō *(or suicide attack) missions. This is more commonly known as the* "kamikaze" *by western people. As a college student, I read with fascination about Japanese Navy pilot Lieutenant Morimasa Yunokawa, and his group, the elite 721* Ku *(the* Jinrai Butai *or Divine Thunder Unit).*

During the final year of the war in the Pacific, this group was equipped with totally unique weapon to take the fight to the Allies in a new way – the Yokosuka MXY-7 Ohka *(or "cherry blossom") rocket-powered, attack plane. This small aircraft would have a 2,646-pound (1,200-kg.) bomb in the nose and would be purposely built to do suicide attack missions with no provisions for the pilot's escape. The* Ohka *would have to be dropped from a "mother ship," Mitsubishi G4M bomber then, after firing its three Type 4 Model 1 Mark 20 solid-fuel rocket motors, race to their target at over 600 miles per hour. However, one of the issues faced by these aviators would be the need to fly their* Ohka*-carrying G4M bombers, which were slow and virtually unprotected, very close to their potential target such as an aircraft carrier. Because of this, these Japanese planes would be easy targets for fighter planes or defending ships below.*

During the several times that I met Yunokawa-san, I found him to be very sharp, well-spoken and proud of his days as a military man. I felt very fortunate to being able to meet this man who had participated in so much history.

Background

I was born in 1921, in Sasebo, Nagasaki Prefecture. My father, Chuichi, was a captain in the Imperial Japanese Navy.[1] Our family was from Yunezawa City, in the Yamaguchi Prefecture. But, we couldn't stay there because my dad, while serving in the Navy, was constantly being transferred around to various port cities such as Yokosuka, Kure and Sasebo. On 12 April 1944, my father died of pneumonia. I had one older brother, Tadanari, and two sisters. Since he was the eldest son in my family, Tadanari had to stay in Yunezawa City and attend school there. After this, my brother was drafted into the Army, but in 1945, he was killed during combat operations in the Philippines. As a family anecdote, in 1966, my younger sister, Reiko Yukawa (whose given name Kazuko Yunokawa) was the emcee for the Beatles when they came to Japan to perform.[2]

Navy Days

In the fall of 1939, I joined the Navy, as a member of the 71st Class, at the Naval Academy at Eta Jima. Then, on 14 November 1942, I graduated from the academy. After all of my training was completed, my first assignment was on the battleship *Ise*, working as torpedo specialist. Next, in early 1943, I served on the light cruiser *Agano*. Our ship took part in the Japanese troop evacuation operations from Guadalcanal. We went three times, back and forth to the besieged island, to rescue our men stationed there. It was during this time that I began to feel that the only way to fight back against such a tenacious enemy was to become a pilot so I really wanted to learn how to fly a plane.

Beginning in June 1943, I did my flight training (as a member of the 40th Class) at Kasumigaura Air Base and flew using the Kawanishi Type 93 biplane trainer, which we called the "Red Dragonfly." After completing this initial course, in February 1944, I went to Ōita Air Base where I learned how to fly fighter aircraft. Then, at the end of April of that same year, our entire group went to Tsukuba Air Base.

The *Jinrai Butai*

In August 1944, I volunteered to become a member of a *tokkō* unit, but I wasn't "officially" part of this operation until the 6 November of that same year when I arrived at Kōnoike Air Base as a member of the 721 Ku (the *Jinrai Butai*). Initially, there was no "pressure" to join; this occurred much later on. On 10 December 1944, I became the *buntaichō* of the #3 *Ohka buntai* with fifty-three men working under me. Sadly, only five of my men survived the war; two of these airmen were injured and couldn't fly anymore. This probably saved their lives.

A lot of people wanted to be a part of the *Jinrai Butai*, but only two-hundred men were selected to join this group. I recall we had a lot of visitors to our base which ranged from high-ranking Japanese Navy officers to *sumō* wrestlers and theatre groups. We also got a lot of gifts from these visitors such as commerative daggers, *hachimaki* and cigarettes. Local villagers also would bring delicious sweet potatoes for us to eat. You could really *feel* that everyone had such high expectations for us (to succeed in the war).

In the *Jinrai Butai*, there were two commanding officers of the Mistubishi G4M bomber units in our group, Lieutenant Commanders Gorō Nonaka (711 *Ku*) and Jirō Adachi (708 *Ku*). Both men were superb leaders and very experienced aviators who participated in numerous combat engagements. Nonaka-san was in our group since it was formed on 1 November 1944. Adachi-san came a little later that same month from another detachment. Because of this, initially, I was able to spend a lot of time with Lieutenant Commander Nonaka. On several occasions, we had drinks together and I could hear some of his stories firsthand. Nonaka-san repeatedly said that he didn't want to do combat missions indiscriminately. Instead, he felt that it was more prudent and less people would die if we did our attacks at night. I recall him saying this many times in front of several of our men and it was "hard" for me to hear this. However, Lieutenant Commander Nonaka had a lot of success doing nighttime missions in the past; he knew how difficult it was to attack during the daytime hours.

Nonaka-san was well-known for making and drinking Japanese tea with his comrades; this was very unique for a person like him (a Navy officer) to do. I thought this guy was a very fun person to be with. He wasn't concerned with the minute details of doing a mission and used a lot of gestures while speaking to you. Instead, this seasoned naval aviator was good at doing the actual combat and getting the job done. He was very brave but had a calm personality. Before flying a mission, many other officers did long speeches; Nonaka-san was totally different in this regard. For example, he would say, "I will tell you our plan when we get there, just get on the aircraft." Nevertheless, this veteran was still able to be very successful in combat on numerous occasions. As an anecdote, In December 1944, Lieutenant (junior grade) Shoichi Ota came to my quarters to complain about Lieutenant Commander Gorō Nonaka, who was in charge of his respective group. Lieutenant (junior grade) Ota was trying to say that his leader was a coward and simply procrastinating from flying a mission using the *Ohka*. However, I knew Nonaka-san wasn't this type of person so I scolded him. I believe I am one of the few people to have met him after the war; we met in 1951.

There were three other *Ohka* unit *buntaichōs* besides myself, Lieutenant Akira Hirano, Lieutenant Kentarō Mitsuhashi and Fujio Hayashi. After Mitsuhashi-san was killed during the first *Ohka* mission on 21 March 1945, I became the overall leader of all these *buntais* including the surviving men from Mitsuhashi's group, Hayashi's men and everyone else. Therefore, on 24 March, the entire *Jinrai Butai* had to be reorganized. Since so many G4M bombers were lost on this earlier mission we had no mother aircraft to fly our *Ohkas* into combat so the decision was made to use our Type 0 carrier fighters equipped with 500-kilogram bombs.

Lieutenant Commander Gorō Nonaka was a legendary pilot in the Imperial Japanese Navy. Being the leader of the ill-fated first Ohka *mission on 21 March 1945, reaffirmed his place in Japan's wartime history. (Morimasa Yunokawa Collection)*

Lieutenant Commander Gorō Nonaka's men at Kanoya Air Base, just prior to take off on their last mission, 21 March 1945. (USN)

The first Ohka mission, on 21 Match 1945, was a complete disaster. All eighteen G4M bombers (of which, fifteen were equipped with Ohkas), were led by the illustrious combat-hardened Navy aviator, Lieutenant Commander Gorō Nonaka. They took off one after another from Kanoya Air Base. Prior to flying this mission, Nonanka-san, felt this would be his last mission but still proceeded with it as ordered. Every G4M bomber was lost without scoring a single hit on their intended target, Task Group 58.1. This respective group of American Navy ships included the carriers Hornet, Bennington and Wasp. Sailing with these respective vessels were the light carrier Belleau Wood, escorted by the battleships Indiana and Massachusetts along with a large supporting group of cruisers and destroyers. The Japanese G4Ms and their lethal Ohkas were supposed to have been escorted by fifty-five Type 0 fighters with thirty-two of them coming from the 721 Ku (flying from Kanoya) and twenty-three more from the 203 Ku, based at Kasanohara Air Base. Out of the thirty-two Zeros from Kanoya, one crashed during take off, and another twelve of them returned back to their base due to mechanical issues on their aircraft. An additional twelve Type 0s from the 203 Ku flew back to Kasanohara because of problems with their fighters. So, in the end, only thirty escort fighters flew with the G4M bombers to participate in the attack on Task Group 58.1, but it didn't make a difference. Out of the remaining Zeros, ten 721 Ku aircraft were lost in action including the leader Lieutenant Mutsuo Urushiyama. Another two Type 0s from the 203 Ku crash-landed because of combat damage to the aircraft. This mission was a terrible loss for the Jinrai Butai with 163 men perishing in the first attempt to use the Ohka in combat.[3]

Above: Picture taken in March 1945 at Usa Air Base, Kyūshū. Yunokawa-san is in the front left. Behind him is Lieutenant Yoshio Onizuka and bottom right, Lieutenant Kanehikō Narita. All three men were members of the 71st Class from the Naval Academy. Lieutenant Narita died on 6 April 1945 while doing a tokkō *mission. (Morimasa Yunokawa Collection)*

Below: This image is from April 1945 at Ōita Air Base, Kyūshū. Left to right: PO1c Nobuo Dodo, CPO Mitsuharu Fujishiro, Lieutenant Morimasa Yunokawa, CPO Yoshiharu Domoto and CPO Kenichi Sato. (Morimasa Yunokawa Collection)

A preserved Yokosuka MXY7-K1 Ohka *trainer on display at the National Museum of the United States Air Force. Note: the wooden skids for training purposes. (USAF)*

The very rudimentary cockpit of the surviving K-1 Ohka *trainer. (USAF)*

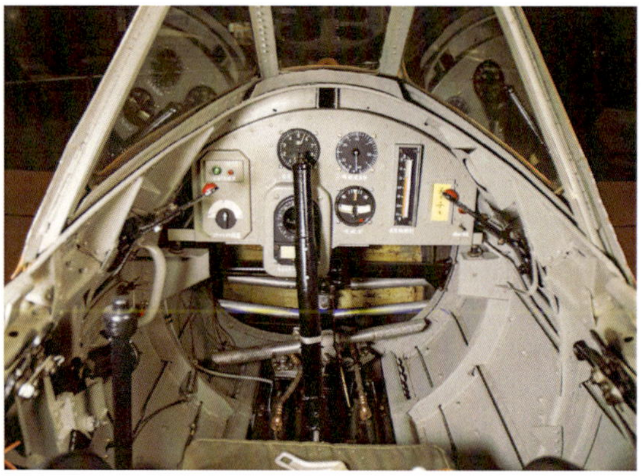

Right: For comparison purposes, an operational MXY7 Ohka *(Model 11) captured during the Battle of Okinawa, but the image isn't dated. (USN)*

Ohka training

All of the *Ohka* training was done at Kōnoike Air Base because it had one large runway and an additional landing strip of sand that was 2,300 meters long. The latter landing strip was not really a runway per se because it was really just sand. However, this second strip was perfect for landing practice with our *Ohka* trainer, which had a skid mounted under the fuselage specifically for this purpose. After a trainee completed the entire practice program, his final test would be done at Hyakurihara Air Base (Ibaraki Prefecture).

The *Ohka* was very difficult to fly. This is because it had no wheels to land. This aircraft had much shorter wings than a typical plane and was heavy in the nose. Initially, we mounted a cushion-type head protector in the cockpit but it was only used a couple of times (in training). After I was the sixth person to land the *Ohka* safely, this head protector was never used again. I recall a few of the trainees that bounced up and down very hard when they didn't land properly. This training was very dangerous and there were three accidents, which resulted in two deaths and an additional person who was injured. Keep in mind, each *Ohka* had to be dropped from a mother ship (G4M bombers) and speed when it was dropped was very fast, so, it could be very lethal as well.

*****

For training with the Ohka, *of course, there would be no heavy bomb in the nose or rocket engines in this small aircraft so in order to compensate for this, water ballast was added to the unpowered trainer.*

Above: According to Yunokawa-san, this image was taken 8 April 1945 at Tomitaka Air Base. It was a farewell ceremony for aviators that were part of Lieutenant (junior grade) Hirokazu Ushikubo's group. The officer in the front center of the picture is Vice-Commander Kunihirō Iwaki. Yunokawa-san is wearing a flight suit and gear and looking at the saké tray while the person behind him (to the left) is Reserve Lieutenant (junior grade) Hachirō Hosokawa. On 16 April (of that same year), Lieutenant (junior grade) Ushikubo was killed in a take off accident. (Morimasa Yunokawa Collection)

Right: Ohka pilots just before their mission. Note that behind them is a Mitsubishi G4M bomber with a lethal Ohka *mounted underneath it. (Morimasa Yunokawa Collection)*

A group picture taken in May 1945 at Tomitaka Air Base. Most of the men are members of Fighter 306 Hikōtai. *Lieutenant Yunokawa is sitting in the center of the first row. To the right of him is Lieutenant Akae, a mechanic* buntaichō *for Fighter 306* Hikōtai. *(Morimasa Yunokawa Collection)*

Thoughts on the Use of *Tokkō* Missions

I THINK WESTERN PEOPLE cannot understand Japan's *tokkō* missions we did during the Pacific War. This is because they don't know the details of the dire situation that our country was facing at that time. Our *tokkō* units were formed as a result of what was happening in the war. Until the summer of 1944, higher-ranking officers tried to contain the feelings of their subordinates, but this changed after our defeats at Guadalcanal and the Battle of the Philippine Sea. After these losses, many outspoken junior-ranking Navy men suggested doing these suicide missions; then, later on, this concept was adopted as official policy of our service. Therefore, in August 1944, an application system was created and during October, it was implemented.

Personally, I believe that most of my men thought about strategy such as the best way to hit a target and sink the enemies' ships. They couldn't wait to fly as soon as possible; at least I thought so. In fact, there was one pilot named PO1c Yukiyasu Fujita who was asked to participate in a mission when he wasn't feeling well. He was asked to join this combat sortie because one of the assigned participants had been injured. PO1c Fujita flew this mission and never came back. Remarkably, this was his fourth attempt to do a suicide attack with an *Ohka* but on the three previous attempts he had returned to our base without completing his mission.[4] I would think that on the previous attempts, Fujita-san returned because a suitable target couldn't be found so he returned, not wanting to squander his life. As I far as I know, I think, most of our men wanted to go and fly these suicide attack missions.

As an anecdote, the overall planning for any missions would always be done by one of my superior officers. After these respective higher-ranking officers determined what the enemy was doing, they sent a message to me to prepare for a mission. Then, after I received this important information, I selected the men who would be flying that day. However, none of this would be determined until that moment because I couldn't know beforehand how many people would be needed because the situation was always changing (during the war).

The End of the War

ON 15 AUGUST 1945, I was at Komatsu Air Base (in Ishikawa Prefecture). Earlier a lot of our men had been stationed in Kyūshū but were now at Komatsu. That morning, we heard that there was going to be an announcement from the Emperor. Then, at noon, we lined up to listen to his radio broadcast but it was difficult to hear. So, I tried to use a different radio so I could hear his message better, but still had trouble hearing it. However, as I looked at the expressions of everyone around me and I thought, something strange must be happening. But, I still wasn't sure what was happening because just before this announcement, the Soviet Union had declared war on Japan. Therefore, a lot of people thought that this message from the Emperor was going to tell us to fight harder, which is also what I thought it could be about. However, this all changed when I happened to be talking with my buddy Lieutenant Commander Hisashi Watanabe. He was the son-in-law of The Naval Affairs Chief, Vice-Admiral Zenshirō Hoshina. Just by chance, I was able to read Vice-Admiral Hoshina's letter to his daughter which explained in great detail the Emperor's decision to end the war. The Vice-Admiral was concerned about his daughter and didn't want her to commit suicide. After I read this letter, I finally was able to understand that the war was over and it made me very upset and stressed out.

*****

ACCORDING TO YUNOKAWA-SAN's detailed wartime records, the Jinrai Butai *(721* Ku*) lost 430 men during the war. This figure includes everyone in the unit from* Zero *escort pilots to* Ohka *mission participants. For their sacrifice, this unit only sunk one non-capital ship, the destroyer,* Mannert L. Abele *(DD-733) and damaged six other American vessels.*[5]

Postwar

JUST AFTER THE WAR ended, I went "underground" and even changed my family name to Ishimura. I recall that all together maybe six or seven of us (Navy officers) went underground because we didn't know what would happen to us after the war was over. However, in early 1946, I changed my name back to Yunokawa. Then, I got a job doing mine sweeping which was still a part of the Japanese Navy after the war ended. At that time, there were thousands of mines around the Japanese mainland. Next, in 1954, I joined The Japanese Air Self-Defense Force (JASDF) and served until July 1976 when I retired as a major general. While I served in the JASDF, I was able to meet many of American Naval officers and they were all very kind to me, which impressed me. My wife and I have two sons and one daughter.

Reflections

During World War II, I think that it was only natural for us to fight for Japan; just like the Americans who fought for the United States. At that time, I thought, the Japanese people might be eliminated and our country would collapse if we didn't fight the ABCD countries (Americans, British, Chinese and Dutch). These western nations stopped selling Japan iron ore, oil and steel; this made life for us very hard. So, in my opinion, the war was really an act of self-defense and the decision was made to attack Pearl Harbor. I was very happy to get the opportunity to fight against our enemies. However, my father disagreed with my opinion and thought that we would have a difficult "fight" ahead of us. This really surprised me; but, some of my counterparts in the Navy also felt very reluctant to have started this war.

The author during one of his visits with Yunokawa-san in 2002. (Ron Werneth)

[1] Captain Chuiichi Yunokawa's cousin was Reiko Mihashi, who later married Admiral Isoroku Yamamoto.

[2] In postwar Japan, Reiko Yukawa was a well-known music critic, disk jockey and even translated lyrics for English language music. She is still active in the music industry today.

[3] The background information for this ill-fated mission on 21 March 1945 was gathered from numerous Japanese-language sources and www.combinedfleet.com

[4] PO1c Fujita was killed in action on 11 May 1945, but nobody saw his final moments so it isn't clear how he died. There was no radio communication with his aircraft after take off. The young aviator took off as an *Ohka* pilot of one G4M bomber (mother ship), with three other bombers. One of the G4M bombers had engine trouble so it returned back to base and did a crash landing. Compendium book of stories authored by veterans of the 721 *Ku*, 722 *Ku* and K708, *Shogen, Ohka Tokko, Ningen bakudan to yobarete*, 574-575

[5] http://www.asahi-net.or.jp/~un3k-mn/ooka-history.htm

Chapter 19

"Why did so many of my comrades lose their lives and I survived?"

Lieutenant (junior grade) Saburō Abe (IJN)

I FIRST HEARD ABOUT Saburō Abe at a reunion of former Zero *fighter pilots from a mutual veteran friend. My buddy thought that Abe-san would be an interesting and important person to interview because he fought in one of the last air battles of the Pacific War on 15 August 1945, just a few hours before the Emperor's official surrender announcement. Saburō Abe also flew with one of Japan's most famous Navy wartime aces – Lieutenant (junior grade) Tetsuzō Iwamoto. Abe-san is very well-read, intelligent and authored several books about the Imperial Japanese Navy and aviation history in his postwar days.*

Early Days

I WAS BORN ON 20 August 1923 in Ashio Town, Tochigi Prefecture. My town is famous for a copper mine incident, which was the first industrial-related problem in Japanese history, where copper poisoning occurred. This well-publicized event happened in the early Meiji Period and it made my town became well-known throughout Japan.

My father worked as an engineer in the Ashio Copper Mine. I was the third boy of six brothers; hence, my name was Saburō, which meant third. Four of us ended up serving in the military and went to war. We have two sisters in our family. I joined the Imperial Japanese Navy because of two reasons. First, at that time, there was a social movement in Japan for all young men to serve in the military. Early on in life, I wanted to be a carpenter but later I chose to join the Navy. Second, I was one of eight children in our family. My dad insisted that all of us should go to college. In Ashio, there was only a primary school. If you wanted to attend junior high school, you would have to go somewhere else such as Maebashi or Urawa. This would cost 15 yen per boy, each month. It would be five yen for school tuition, another five yen for accommodations and still five more yen for food/expenses. In those days, very few working men earned over 100 yen per month. However, since my father was successful engineer, he probably earned more than this per month. Nevertheless, it would be very difficult for him to send us all to junior high school.

*****

Even though it was difficult for Abe's father to pay the cost for so many children to attend higher-level education, Saburō Abe still went on to Maebashi and Nakano junior high schools before applying for admission to the prestigious Naval Academy.

Abe-san Joins the Navy

Frankly speaking, I didn't want to join the Army. I was trying to decide if I wanted to join either the traditional Japanese Imperial Navy or become a member of our Merchant Marine force. After discussing my options with my father, he quickly replied that service in our Navy would be better for me, so that's what I did. After joining this military service, I became a member of the 73rd Class at the Naval Academy in Eta Jima. There, I saw the possibility of becoming an aviator and thought this was a better option than serving on a ship.

In March 1944, after graduating from the Naval Academy, I became a member of the 42nd Pilot Training class, and went to Kasumigaura Air Base for my primary instruction. This course lasted until August of that year. Next, in September 1944, I went to Tsukuba Air Base and did my advanced training; flying the *Zero* fighter. In February 1945, after completing all of my flight training, my first assignment was with the 203 *Ku*, 311 *Hikotai*, at Kokubu Air Base, Kyūshū.

Abe-san pictured with a Zero fighter. (Saburō Abe Collection)

Fighter pilot Saburō Abe in his typical flying garb. (Saburō Abe Collection)

Off to the Frontlines

On 19 March 1945, I arrived at Kokubu Air Base, but the day before it was devastated by planes from the United States Navy (Task Force 58). The base was in absolute chaos and there wasn't much for me to do. Therefore, soon after, I transferred to Izumi Air Base, Kyūshū. While I was at Izumi, there was about forty of us (pilots). Even though I was designated to be a *shotaichō*, I only had about one hundred hours worth of flying time. Whereas, my wingman, legendary fighter pilot ace, Lieutenant (junior grade) Tetsuzō Iwamoto had well over 1,000 hours of flying time. This amazing aviator had rose through the ranks as an enlisted man while I was a 'green' officer-rank pilot who just went to the Naval Academy. Iwamoto-san seemed to like me and he said as long as we flew together, I wouldn't be killed.

Abe-san snapped this picture of a Kawanishi Shiden *(Type 11) fighter during his training days at Tsukuba Air Base in the fall of 1944. (Saburō Abe Collection)*

During this time period, I only flew the A6M5 *Zero* fighter, (Model 52). Really, the only missions I flew were to protect our *Suisei* (Allied code name "Judy") dive-bombers, which had the air-cooled engines. Remarkably, these dive-bombers were faster than our *Zero* fighters so we had to take off earlier than them and then fly to Okinawa, where we circled around waiting for them to arrive. However, the United States utilized excellent radar so they always knew when we were coming.

Lieutenant (junior grade) Tetsuzō Iwamoto Up Close

Lieutenant (junior grade) Tetsuzō Iwamoto was the top-scoring ace of the Imperial Japanese Navy, with an astounding self-estimated kills to be 202 downed enemy aircraft. However, postwar historians estimate that he may have scored around eighty aerial kills. This remarkable aviator rose from the enlisted ranks as a typical sailor to become an eight-year combat veteran and top Navy ace. Iwamato-san's amazing career spanned from 1938-1945 and he flew in numerous engagements from mainland China to being in the Solomon Islands, and eventually, the final defense of the Japanese mainland. During the spring of 1945, the subject of this story, Saburō Abe, flew with him during his days with the 203 Ku, *311* Hikotai. *Sadly, after having survived so many years in combat and the ensuing chaos that occurred at the end of the war, Iwamoto-san only lived ten more years. In 1955, he died at the age of thirty-eight years old after surgery to alleviate some war-related back wounds.*

*****

Lieutenant (junior grade) Tetsuzō Iwamoto-san was full of life and always had lots of energy. At night, there was nothing to do so he brought a 1.8 litter (about half-a-gallon) bottle of *saké* for us to drink together. Typically, nowadays, it is polite to pour liquor for your drinking companion and vice-versa. However, back then, when we

Lieutenant (junior grade) Tetsuzō Iwamoto was a legendary ace in the Imperial Japanese Navy. He befriended Saburō Abe, at that time a rookie fighter pilot, while they served together in the 203 Ku, 311 Hikotai. *(Saburō Abe Collection)*

drank together, we didn't do that. Instead, we each had a big cup and poured our own drinks. After we finished drinking, we got on a motorcycle with a sidecar. At that time, I was a *shotaichō*, and technically, out-ranked him. Nevertheless, Iwamoto-san made me 'drive' and he rode in the sidecar! Then, we went to the prostitutes' quarters together. It seemed like almost every night, he picked a different prostitute to sleep with. In my case, I was still a virgin, so I didn't do that. Instead, the madam of this house treated me to *saké*, but it was very stinky. I really didn't know much about alcohol but this was really *shōchū* (made from sweet potatoes). Then, after Iwamoto-san finished up, we went back to our quarters. Typically, the following morning (after these nights), this amazing fellow would say, 'Let's go flying!'

Interestingly, Lieutenant (junior grade) Iwamoto may have not had 'better' eyesight than his fellow airmen but he was always first in spotting enemy aircraft. So, one day, I asked him, 'How are you able to find enemy aircraft so quickly?' His blunt reply, 'Well, this doesn't depend on my eyesight; I use my common sense. Think about it for a minute. The enemy aircraft would be guided by radar so they would be waiting for us to come to the area. So, while they were doing this, the enemy pilots would have to fly in circles. Therefore, sunlight would reflect on their aircraft and you would be able to see some reflection. You had to be in the same position as your American counterpart. If 100+ enemy aircraft were flying in circles, there would always be some reflection.' Iwamoto-san went on to say that if I always looked around for them in 360 degrees, it still would be impossible to find the enemy unless you used his method.

After Lieutenant (junior grade) Iwamoto located the enemy aircraft, he quickly pointed them out to the group leader; then only he and I would go after them (before everyone else). Iwamoto-san was like a 'brother' to me. The two of us would get some altitude then dive down and commence our attacks. So, generally, we scored more hits than the other pilots in our unit. Our squadron-mates always operated as a 'group' – based on what the group leader said. This would limit their success in air combat. The other pilots complained that we were sneaky but this was Iwamoto-san's style. We used hand signals because our radios didn't work well. Today, people probably think air battles last for a longer period of time, for example, twenty minutes or longer. In reality, though, it would be over in just a few minutes. Earlier in the war, our motto was, 'One shot, one kill,' but, in those days, it was 'One shot, one escape.' However, since his early days in combat, Iwamoto-san's technique was always one shot, then escape; and repeat it again. This was new to me, because my flight instructors always said that you should stay with your group and continue fighting together. Keep in mind, this veteran's technique was not any official policy; it was his own 'style.' The technique also was largely dependant on the performance of our aircraft. During the late stage of the war, it was impractical to continue fighting. I thought air combat was really hard to continue once you engaged the enemy then lost altitude and acceleration. Just to be in a position to initiate air combat, you had to be very lucky.

Lieutenant (junior grade) Iwamoto would get some altitude then swoop down on his 'target' aircraft while firing at it. He was very skilled at gunnery so his success rate was very high. Iwamoto-san would always lead me to a good position to commence an attack on our opponents. For me, a target would always be there; it was just a matter of pulling the trigger. To be frank, he did everything for me – all I had to do was pull the trigger. This amazing aviator was a very innovative person. Typically, at Rabaul, our pilots would try to shoot down the enemy bombers before they dropped their bombs. Iwamoto-san was different. Instead, this ace waited for them and after they dropped their bombs, he would try to shoot them down. He did purposely because enemy pilots would be probably 'scared' after completing their mission while trying to flee the area. Before they did their attacks, they would be very focused and determined to complete their duties. Thus, this ace purposely tried to target the enemy airmen after they were not as focused and ready to get home.

Abe-san in the cockpit of his Zero *fighter, (Model 52). (Saburō Abe Collection)*

Chaotic Spring of 1945

THE NIGHT BEFORE PLANNED combat missions, the names of the selected pilots and their designated time slots were written on the roster board. For example, at 0800, these men would fly, and then at 1000, such and such were flying, etc. Based on this predetermined time table, if the enemy fleet was found at 0800, a *tokkō* mission would be done. If, for example, enemy aircraft came, we would scramble and intercept them. Everything was done on a case by case basis, we would only be assigned a time. It was a very chaotic situation. Fortunately, for me, on my designated time slots to fly, there were no *tokkō* missions because we never located the enemy fleet. As an anecdote, I recall four designated *tokkō* missions were done while I served with this group.

It must sound strange but I don't recall any of the names of my immediate superior and commanding officers at that time; everything was so unorganized. I recall back in 1970, while playing golf with a group of former veterans, I told a gentleman that I was at Kokubu Air Base at this time period in 1945 and he said, 'Ah, I was probably you're *buntaichō*!' That's how crazy things were in the last year of the war.

*****

Around this time period, in early April 1945, Abe-san and his comrades were expecting to fly aerial cover for the battleship Yamato *but they never got the opportunity to even take off since the once proud ship was sunk on 7 April during the ill-fated journey to attack Allied forces off Okinawa.*

*****

A few days after the battleship *Yamato* sank my superior officer said that we were 'useless' and should go somewhere else. Then, we transferred to Mobara Air Base, Chiba Prefecture. Around 10 April 1945, I boarded a Type 96 transport aircraft (converted from an old bomber) and there wasn't even enough time for me to gather up my belongings. I complained but didn't have a choice in the matter and we flew to Mobara.

Abe-san's Last Assignment in the War

My next assignment was with the 252 *Ku*, 304 *Hikotai*. Initially, I went to Mobara, Chiba, but due to the concern of being attacked by enemy aircraft, our group went to Kuriyama Air Base. I was stationed there when a B-29 attack on Sendai occurred but I wasn't supposed to fly that day after just arriving at my new base. In total, eight aircraft scrambled to intercept them. I took off later, but by the time I approached Sendai, they were already gone. So, I circled around and went back. In the end, only one of our fighters spotted the B-29s because they were so fast. We simply couldn't catch them.

On another occasion, I was assigned to fly an escort mission for small convoy of four freighter ships which were traveling from the Kujūkuri area to Cape Inubō. A group of enemy dive-bombers attacked at a very low-altitude (much lower than we estimated) but scored no hits. This was another failure on our side but without a precise briefing or more information, how could we be successful?

Even a few months earlier, during missions over Okinawa; I figured that I would be dead within a year or so. I already had felt that we couldn't win this war. Since Japanese Navy pilots were considered to be 'elite,' we always had a carefree attitude even if we knew that it would be a suicide mission. In the 252 *Ku*, I was one of the youngest aviators with the least amount of flying time compared to some of the other men who had been pilots since the early days of the war.

Another 'memorable' mission for us when the pilots of my unit sank an enemy submarine (but don't have the exact date). For this sortie, four airmen were waiting on standby and would take off if enemy activity was reported. After a report came in for an enemy submarine, they scrambled and went to try to sink it. Remarkably, a very inexperienced pilot fired upon on the area behind the submarine conning tower then it exploded and sank. He must have scored a hit with his 20mm cannons on ammunition storage area, which blew up. We all celebrated this remarkable feat by a young aviator from our unit.

The Loss of a Good Friend

A dear friend of mine was Lieutenant Yoshiō Saitō; he was a 71st term graduate of the Naval Academy and a *buntaichō*. In early August 1945, B-25 bombers came to our area, almost every day, at a certain time, just like clockwork. This incident occurred on the 5th of that month, while we were at Mobara. On the previous evening, Saitō-san and I had an enjoyable night together drinking. So, on the following day, we were in the pilots' ready room talking about how much fun we had together. Then, an alert came in that our radar discovered a medium-sized enemy aircraft. This was probably a B-25 and it was flying from the north, the Katsura Beach area. After the warning came in, Lieutenant Saitō said, "It is my turn this time," and then hurriedly took off in his *Zero* fighter (Type 52) to intercept this plane. Sadly, he was shot down soon after taking off. Lieutenant Saitō was like a brother to me and I dreamed about him several times after his tragic death.

A picture from Abe's personal collection showing a group of 252 Ku, *304* Hikotai's *A6M5* Zero *fighters getting ready to depart for a mission from Kuriyama Air Base. (Saburō Abe Collection)*

Air Combat - 15 August 1945

Noted Japanese historian Ikuhikō Hata comprehensively covers the air action over Japan on 15 August 1945 in his book including the following aerial engagement that Abe-san, as a member of the 252 Ku, 304 Hikotai, *shares below.*[1] *The leader of Abe's* hikotai *was Lieutenant Commander Moriyasu Hidaka and the overall commanding officer of the* 252 Ku *was Commander Kiyoji Sakakibara. On this date, at 0400, the men of 252nd were woken up. Both the 304 and 316* hikotais *scrambled forty* Zero, *(Model 52) fighters apiece to engage an incoming enemy carrier strike. At 0540, the Japanese defenders were ordered to take off to engage this large group of strike aircraft. Take off conditions were very poor with thick fog and bad visibility but the* Zeros *still launched as planned. After joining up into flying formation at 6,000 meters (19,685 feet), they spotted a huge group of enemy aircraft which was a mixture of both naval bomber-type aircraft and escorting fighters. Abe-san continues with the rest of his experiences including this mission, his last, that fateful day.*

*****

On 15 August 1945, the 252 *Ku* really had no advance notice that this would be the day the war officially ended. However, some of the other Navy units such as the 302 *Ku* already knew a few days earlier that on the 15th, our Emperor would do an official surrender announcement. For the 252 *Ku*, it would be combat operations as usual. On the day (the 14th) prior to the surrender, the enemy fleet (*Indefatigable* from the Royal Navy and the United States Navy carriers *Essex*, *Randolph* and *Wasp*) were in the waters off Chōsei Beach, Chiba. We took off to try to attack them but it was nothing but dense fog and thick clouds. So, after taking off, I could only see the tops of trees and above them was nothing but clouds. So, the plan was to try again, early the following morning.

Early the next morning, a mechanic came to my assigned fighter and started filling it up with fuel. Because of the heavy mist that day, I couldn't even see the control tower so I had to listen to the instructions from my radio. In order to do that, I had to turn on a main switch, which powered up not only the radios, but other systems such as the guns. As I was listening to the radio with headphones, I stood on one of the wings, while the mechanic filled up the fuel tanks. While this man was fueling my *Zero* fighter up, another ground crew person was in the cockpit, readying the aircraft for this mission. Tragically, while this second guy was doing this, he accidentally touched the trigger, which controlled the guns and bullets started to fire. Then, the poor mechanic was killed as he was fueling my plane up. As this man died, his eyes met mine as he probably thought I was responsible for his death. I felt so bad after this terrible accident happened; it was horrible.

After this tragedy, I didn't want to fly this fighter, which was covered with blood and needed to be cleaned up. So, I ended up taking another plane and took off later than the main group of our aircraft. This was in the early morning of 15 August, before the sunrise. Since I was the last person in the air, I tried to hear what Lieutenant (junior grade) Yoshi Tanaka was saying over the radio. The air battle was already underway as I was trying to catch up. For some reason, instead of targeting the lead enemy aircraft, Lieutenant (junior grade) Tanaka, my *taichō*, went after the last one (or 'Tail-end Charlie') in the formation. Because of this, our pilots didn't know what to do; it was utter confusion in the air. When they started the engagement, our *Zero* fighters were probably about 500 meters (1,640 feet) above the enemy aircraft and in a better position to attack them. Tanaka-san was leading two more of aircraft against a group of enemy planes that were probably four in number. While I was listening to all of this, I could clearly see what was happening. Since one of the enemy aircraft was banking, I vividly saw the pilot looking over his shoulder directly at me. During my gunnery training, I was only good at strafing targets on the ground while my air-to-air attacks were very poor. Because of this, I committed myself to get very close to my 'target' plane before commencing my attack. As I gazed at him, my opponent looked so scared and, in that instant, he must have realized his mistake. Then, I pushed down my trigger and fired all my guns at him. The enemy pilot's head was blown apart and his blood covered the inside of the canopy glass. His plane then gained some altitude and luckily, I managed to pass under my victim.

As I was flying away from the scene of air battle, two of the comrades of the fallen enemy airman began to chase after me. Just then, the sun started to rise and it was behind them. They were flying single, low-wing, fighters and they didn't look like the typical Grumman aircraft that I was familiar with. I thought their wings looked 'longer' than the usual enemy planes. Could they be 'friendly' Japanese aircraft? It seemed like they were about 200 meters (656 feet) away from me so I thought it may be good to waggle my wings to see if they were on my side or not. At that very moment, the image of my good friend Lieutenant Saitō appeared in the windshield of my cockpit and yelled at me, 'Escape, Escape!' This happened just as the enemy aircraft were shooting at me. So, I tried to evade their attacks by banking my fighter but they scored hits on me and I felt a sharp pain in my leg. I realized that these planes were not 'friendly' when I saw the British roundel markings on them.

Next, I tried to do a counterattack on them, banked my *Zero*, (Model 52), and attempted to gain some altitude. But, oil covered the canopy of my fighter so I slid it back to try to improve my visibility. Then, oil splashed on my goggles so I couldn't see a thing and decided to give up the fight. After I took off my goggles, I thought they would try to attack me again. Sure enough, bullets began to zip past my fighter. I could see the bullets as they fired their guns at me. I returned fire with them but I don't think either side was successful in scoring hits.

After these enemy fighters climbed and flew off, I scanned my instrument panel. I noticed the oil pressure was quickly going down so I flew into the clouds and escaped. The engine seemed like it was just about ready to quit. In the sky around me, I spotted five Japanese pilots and two enemy airmen floating down in their parachutes, but one of them was clearly slumped over, possibly dead. Even though my air battle was over the ocean, by this time, I was already over land. Therefore, I decided to do a forced landing. However, this region was very mountainous in the Chiba Prefecture so I knew that a landing place had to be carefully selected. My engine was barely running when I spotted a river. So, I thought the riverbed along the side of the water would be a good place to land. Thankfully, I managed to avoid the power lines and pulled the control stick way back then turned off the engine.

My landing speed was 136 knots, which I remember to this day. I put my left hand and both legs up against the instrument panel to brace myself for the impact. I shredded a bunch of corn stalks as I did a belly landing just before a railroad bridge. As an anecdote, one side of the river (near my crash-site) was Ichikawa (Chiba Prefecture) while the other side was Koiwa (part of Tokyo). I was worried that my plane would flip over during the landing but it didn't. After this, I removed the vacuum tubes from the radio, took off my goggles and checked my body for wounds. My right leg was in a lot of pain. On that leg, there was a hole towards the bottom of my pants. I also had two more holes in the sole of my flight boot (on that same leg). Luckily, it looked like the bullets just passed through these holes and didn't cause any significant wounds.

Just after I landed, a police officer came from the local station and asked me what happened. I explained that I was shot down and he spotted the blood all over one of my legs. The policeman took me to his station and treated my wounds, which were really superficial. It 'felt' like it was already noon; however, the time was only about two hours after I took off. After I was cleaned up, the police officer put me on a truck to try to get me back to Mobara Air Base. In this vehicle, there was another man, who was an *Asahi* newspaper reporter. He said to me, 'Did you hear that at noon today there would be a radio broadcast from the Emperor?' I replied that I had no idea this would happen. He went on to say that the Emperor would formally announce that Japan would surrender. My reply to this reporter was, 'What did you say?' This scared him and he then explained everything to me. But, I thought the Emperor was going to encourage us to continue fighting after Russia declared war on Japan.

Just as I got off the truck at Mobara, the Emperor's radio announcement started. Once I heard his message, I realized what this reporter had told me was true. So, I thought it was important to get back to my base ASAP. Then, I borrowed a bicycle from a nearby family and quickly rode back there. After arriving at my base, I reported to the *hikochō*, Shingo-san. I explained to him about the air battle, shooting down the one enemy fighter and my crash-landing. That night, on the 15th, I still had lots of pain in my right leg. I got very drunk after drinking so much *saké*. I thought about what is the best way for me to end my life. This thought went through my head over and over until the next morning. The following day, on 16 August 1945, I still was confused and didn't know what to do.

*****

After careful research, historian Jim Sawruk determined it was very possible that the enemy fighter that Abe-san shot down was an American naval one such as an F4U Corsair or F6F Hellcat and not of British origin as Abe-san believed. However, at this point, there is no way to say for sure. This was a mixed, confused aerial melee and the only British fighter lost that day was the Seafire flown by Sub-Lieutenant Fred Hockley. Hockley's tragic fate is discussed later in this account as well as his comrade's damaged Avenger that ditched at sea. There were no other British aircraft lost or damaged on this date. In conclusion, the fighter that Lieutenant (junior grade) Saburō Abe shot down was probably from the American side.

As an anecdote from the Japanese side of the battle, Abe-san's squadron-mate, Lieutenant (junior grade) Tadahikō Honma had to bail out of his stricken Zero *fighter after it was damaged by a British aircraft. He clearly recalled the roundels on the wings that designated the nation of his opponent. As Honma-san floated down in his parachute, he was strafed by Grumman fighters but not injured. He landed with minor injuries and local farmers helped him. After the war, this ex-Japanese fighter pilot was a professor and told his class that he was shot down by a Spitfire but nobody believed him. Many years later, he visited the United Kingdom and went to a British museum where they confirmed all the matching details of this aerial engagement.*

Postwar

After the war was over, I was a farmer for two years. Then, in 1947, I went to Tokyo University and graduated from there four years later. In 1970, by utilizing a computer, I worked with someone at Columbia University to produce English-as-second-language (ESL) education cassette tapes. We tried to promote this learning program throughout the world. Later, I worked for Asahi Broadcasting Corporation and TV Asahi Production Company until my retirement.

Reflections

Losing the war was something that was really hard to accept. Why did so many of my comrades lose their lives and I survived? I was very fortunate to be alive; so I have lots of mixed feelings about what happened. Every year, on 15 August, I go to the Yasukuni Shrine, to comfort the souls of my fellow classmates from the 73rd Class of the Naval Academy who gave their lives during World War II.

In May 2010, Abe-san passed away.

The author and Abe-san after their last meeting together in November 2007, Tokyo. (Yoji Kutsuzawa Photography)

Post Script: The British Perspective of the 15 August 1945 Air Battle

In the early morning hours of 15 August 1945, the British carrier Indefatigable *launched its strike aircraft consisting of four Fireflies (1772 Squadron), six Avengers (820 Squadron) and eight Seafire escort fighters (887 and 894 squadrons). They were a part of a combined American and British carrier strike of 103 aircraft. Sub-Lieutenant Fred Hockley was the overall leader of the five Seafires that designated as close and middle level cover while Sub-Lieutenant Vic Lowden led the trio of fighters at the top escort. The initial British target of this mission was Kisarazu Air Base but due to poor weather conditions, they decided to bomb a nickel processing plant in Kamogawa (Chiba Prefecture) instead.*

Around 0545, as this British group of aircraft proceeded to the secondary target, they were attacked by a large formation of 252 Ku Zero *fighters. Several accounts from the Fleet Air Arm aviators reported that they initially spotted two low-level* Zero *fighters, which were thought to be decoys. Thereafter, the much larger group of Japanese aircraft pounced on the incoming Royal Navy planes. On the initial first pass, Sub-Lieutenant Fred Hockley was shot down, and then the aerial melee continued on.*[2] *After this engagement was over, Sub-Lieutenant Vic Lowden shared the experience of his final combat in World War II –*

"The fighter escort (was) in model position (with a) middle cover of four Seafires 1,000 feet above strike with top escort a further 3,000 feet above them. At time of sighting enemy aircraft, both escorts were from directly over strike too 500 yards astern, and the enemy aircraft accordingly moved from the beam to one-o'clock, closing rapidly so that the four aircraft got into attacking dive out of range of the top escort.

I took my flight of three aircraft at 100 yards abreast spacing to attack four *Zeroes*. Taking the second from the right I signaled the commencement of firing by opening fire at 800 yards, using G.G.S. (Gyro Gun Sight) and cannon alone, on a 10-degree port attack, closing to 450 yards and astern, firing two (two-and-a-half second) bursts. The starboard oleo and then the port oleo legs dropped, and the enemy flamed nicely, going down. My number three was also firing at this aircraft, hitting the cockpit. I therefore took the extreme starboard enemy aircraft, giving a quick stern burst at 300 yards as he dived to starboard. A shallow climb to starboard ensued and I got a slight deflection angle shot at 250 yards and closed to a stern two-second burst at 200 yards when the undercarriage dropped, bits flew off and the enemy aircraft burst into flames and disappeared in an aileron dive at high speed into the cloud. Cannon and machine gun were used together throughout this attack.

I then found one of the original attacking *Zekes* (Allied nick-name for the *Zero* fighter) climbing at 8,000 feet and about 1000 yards from me. I closed to 100 yards at 11,000 feet, kicked on rudder to have a look at the markings, went back astern and fired two (two-second bursts) of machine gun (cannon being finished). Strikes all over the aircraft and the pilot bailed out, his aircraft diving past him and smoking somewhat. Three *Zekes* then attacked me from 15,000 feet, and I finished my machine gun ammo in two short separate head-on attacks at 400 yards before leaving the area hurriedly."[3]

*****

Sub-Lieutenant Randy Kay, flying in his Seafire (NN584), provided his perspective of what happened on the morning of 15 August 1945 –

"I was in the position for middle cover behind the strike 1,000 yards astern, 500 feet above and spotted one bogey 2,000-5,000 feet below strike. Did not engage but maintained station, possible decoy. About twelve aircraft jumped us from above; I put on full power and climbed. A *Zeke* came into my sights and I closed from the port quarter and fired a short burst, observed strikes on the fuselage. (Then I), closed the range to 200 yards astern and fired another burst, observing more strikes. Enemy aircraft then rolled on its back and went into a dive. I did not follow it down for there were more *Zekes* above. I turned to starboard, climbing and spotted another *Zeke* flying across my sight from port. I turned and closed the range to 200 yards, sixty-degrees off and gave a short burst. The tail blew to pieces and the *Zeke* went down in a terrific spin without a tail. He spun into (a) cloud, which was down on the deck. I continued my turn to starboard, for there were more *Zekes* above me. I closed one on the starboard quarter and gave a burst, observing strikes on the fuselage and mainplanes. He became aware of me and turned into my attack and I gave another burst but my guns ran out of ammunition. More strikes along the fuselage in (the) second burst and a trail of white smoke appeared. I did not wait to see what happened for another *Zeke* was on my tail. I climbed to 16,000 feet and set course for base on my own."[4]

*****

AFTER THE AERIAL BATTLE subsided, only Hockley's Seafire was shot down and one of the Avengers (as related below) was damaged. The seven surviving Seafires made their way back to the fleet. Nevertheless, all of the Avengers pressed on, without escorting fighters to complete their attack on the nickel processing plant. All of the English-language publications and books I have read tend to focus on the well-known, but sad tale, of what happened to Sub-Lieutenant Fred Hockley. This airman successfully bailed out from his stricken Seafire fighter but was later executed by Japanese Army soldiers after the noontime Emperor's surrender radio announcement. In September 1947, the two officers responsible for this war crime, Colonel Teiichi Tamura and Major Nobuo Hirano, were hanged. Captain Masazo Fujino, the man who executed Hockley, was given a fifteen-year jail sentence but it is unclear how long of the sentence he actually served.

Remarkably, while researching this air battle, Japanese historian Minoru Kamada, learned about a second British aviator who perished, Sub-Lieutenant John Francis Joseph Andrew Bonass. This twenty-two year old man, from Thornton Heath, Surrey, England, was a crew member on an 820 Squadron Avenger. The other two men on his aircraft (#JZ.482), were pilot, Lieutenant L. Baldwin (RNVR) and turret-gunner, Petty Officer Airman A. Simpson. The latter aviator recalled the furious air battle on the last day of the war –

"I SAW FOUR *ZEKES* COMING in out of the sun from above and felt the impact of the bullets on the aircraft. By the time I had made my switches and had swung the turret the first *Zeke* was already past. The second closed, firing. It came in on a curve from the port quarter and above. I opened deterrent fire at about 700 yards, but he came on and presently presented a non-deflection shot, head-on. I fired in short bursts until he settled down then I fired continuously. Some flames underneath its engine and it began to pour out a considerable quantity of black smoke. Stall turn to port and dived smoking out of sight. I returned fire of third and fourth *Zekes* without being able to observe any affect." [5]

*****

AS A RESULT OF THIS engagement with enemy fighters, the Avenger lost its starboard-side tailplane and elevator. The mid-section of the aircraft (where the observer would sit) also was damaged and a fire started there. After this, Sub-Lieutenant Bonass could have elected to bail out of the damaged plane. However, the two surviving crew members of Avenger (#JZ.482) still continued on to their target and completed the bombing attack. Then, they managed to nurse their damaged aircraft back to a known destroyer position about sixty miles offshore. This Avenger ditched at sea and both Lieutenant L. Baldwin and Petty Officer Airman A.

Simpson were picked up by a friendly vessel.

Fast forward over seventy years later, I combed through the original squadron report written by the two survivors of his crew, the fate or actions of Sub-Lieutenant Bonass are not even mentioned in the text. Through many years of research by Ichirō Kunō, historian, at the Mutsuzawa Choritsu Rekishiminzoku Museum, we were able to find out about the young naval aviator's final moments. Sadly, according to Kuno-san, after bailing out, the parachute of Sub-Lieutenant Bonass failed to open and he fell to earth like a brick. He went on to say that Bonass was found lying on his parachute with no major wounds or bleeding to be seen. However, the deceased airman did have some small wounds on his chin and one leg.

Sub-Lieutenant John Francis Joseph Andrew Bonass was buried at the foot of a nearby mountain. A priest from Kofukuzenji (Buddhist temple) comforted the spirit of Bonass until the 33rd anniversary of his passing. As per the Buddhist tradition, any person, good or evil, who has been deceased for thirty-three years, is able to be reborn as a Buddha. There was a wooden tablet displayed for thirty-three years to honor him. Today, the son of the original priest continues the same duties as his father did at Kofukuzenji. Every year, on 15 August, the anniversary of the airman's death, he comforts the spirit of Bonass. Lastly, Kunō-san heard a few years after the war; the remains of this British airman were moved to the Yokohama War Cemetery and are there for eternal rest.

After this aerial engagement, the Seafire fighter pilots were officially credited with seven aircraft destroyed, three probables and four damaged. The crew of Lieutenant L. Baldwin's Avenger was recognized with one damaged enemy fighter. With the official ceasefire at 0700 in affect, this was last major air battle for the Royal Navy.

The eternal resting place of Sub-Lieutenant Bonass at Yokohama War Cemetery, Japan. (Steve Rogers, The War Graves Photographic Project)

(See pages 256-257 for a painting by UK-based aviation artist Rod Kirkby titled, "Final Combat" depicting the aerial engagement from 15 August 1945).

[1] Ikuhiko Hata, *The Sky of 15 August, The Last of Japan's Air Force.*

[2] David Brown, *The Seafire: The Spitfire that went to Sea*, 127-129. Along with background research done by Minoru Kamada and Jim Sawruk.

[3] Sub-Lieutenant Victor S. Lowden (887 Squadron), *AIR 50/329/1*, 15 August 1945.

[4] Sub-Lieutenant Randall G. Kay (894 Squadron), *AIR 50/330/2*, 15 August 1945.

[5] Petty Officer Airman A. Simpson (820 Squadron), *AIR 50/324/2*, 15 August 1945.

Afterword

In a blink of an eye, time seems to fly by. It is very hard to believe that World War II occurred over seventy years ago. Thankfully, there still is an interest in this massive war that has had long-lasting affects, which can be seen even to this day. Just take a look at the amount of new movies and books that cover this conflict that are still being released. Nevertheless, with the passing of time, the actual participants of World War II along with their children ("Babyboomers") are fading away. As these people pass away and "new" generations replace them, history is often reinterpreted or completely questioned by the general public who may not have the knowledge to properly understand what happened and reanalyze it. For example, the decision to use the atomic bombs on Japan and the justification for dropping them on two Japanese cities. Would Japan have surrendered if these horrific weapons of war were not used? It is only by studying these historic events from both sides that we can gain an unbiased perspective of what really occurred, and then apply the lessons learned from these tragic events to the present.

I have spent over half of my life so far, talking to and recording the stories of our World War II veterans, both in the United States and Japan. I also have tried to read as much as I can about history (in general) and the people who lived it. In my eyes, this has become a life-changing passion. However, for most people, I believe, it is simple as talking with your own family members and people in your community. I wish that I had this opportunity with my own grandfather, but unfortunately, never did. It is vital for everyone to understand history so the same tragedies don't repeat again – "History always repeats itself." Just look at the military aggression occurring both in the mainland of Europe and Asia. It is my hope that if this book even gets one "new" reader interested about our military veterans and World War II, my work has succeeded.

I have done my best to try to ensure the historical accuracy of this book while utilizing both official American and Japanese resources coupled with the veterans' accounts of what happened. However, if readers find something they feel is an error please contact me and it is possible the work will be revised in future editions. Please visit online at: *www.RonWerneth.com* and *Facebook, JapaneseEmpireFall,* for additional resources and multi-media links.

As my good friend Captain Chuck Downey always said "Press On" – whatever your personal mission in life is.

Fine Art and Aircraft Profiles

"Mission Accomplished" by Roy Grinnell

"Final Combat" by Rod Kirkby

A6M3 (Model 32) *Zero* fighter by Michael Claringbould

J1N1-S (Type 11) *Gekkō* fighter by Michael Claringbould

J2M3 *Raiden* (Model 21) by Michael Claringbould

N1K2 *Shiden Kai* (Model 21) by Michael Claringbould

SB2C-3 Helldiver by George Zielinski, A.G.P.

TBM-3 Avenger by George Zielinski, A.G.P.

"Mission Accomplished" by US-based aviation artist Roy Grinnell depicting the events of Sunday, 18 April 1943, over Bougainville Island. These are the final moments of Admiral Isoroku Yamamoto and everyone on board his doomed Mitsubishi G4M1 – Allied code name "*Betty*" – the transport version of the bomber. They were shot down by First Lieutenant Rex Barber while flying his Lockheed P-38G Lightning, "Miss Virginia." This is one of the most important aerial missions of World War II. (See chapters 2 and 9).
For more work by Roy Grinnell visit http://roygrinnell.com

Miss Virginia
147
323

"Final Combat" by UK-based aviation artist Rod Kirkby. The British perspective of one of the last aerial engagements of World War II on the 15 August 1945, Tokyo Bay, Japan. It depicts Sub-Lieutenant Gerry Murphy, based on the HMS *Indefatigable* with 887 Squadron, Fleet Air Arm, in his *Seafire* F. III fighter (serial number NN.212), shooting down an A6M5 *Zero* fighter. The Japanese Navy pilot of the latter aircraft bailed out. For a more detailed account of this mission please see Lieutenant (junior grade) Saburō Abe's account in chapter 19.
For more of Rod Kirkby's work visit www.artist.abelgratis.co.uk

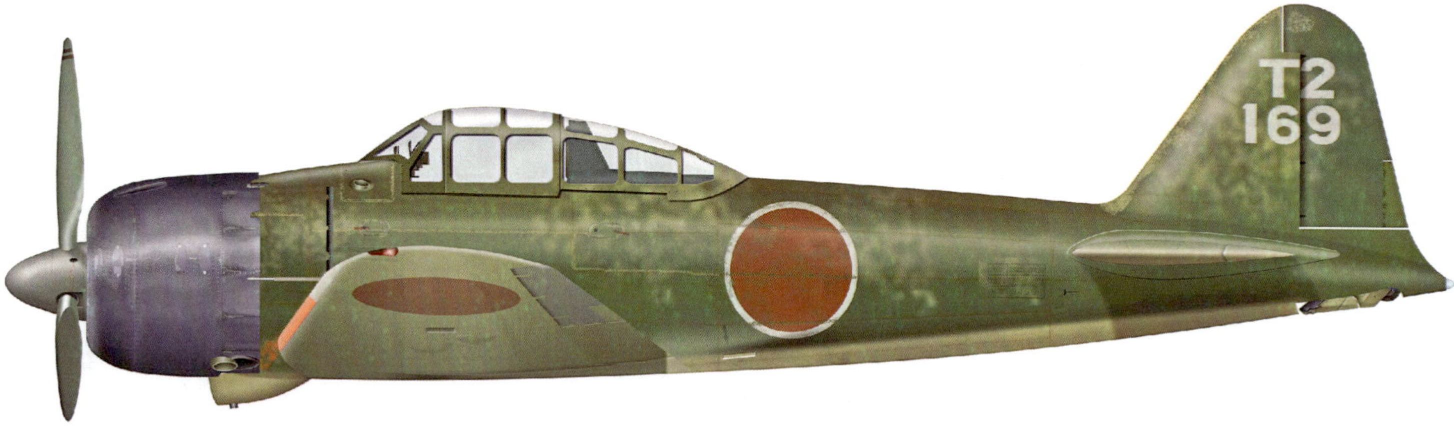

A6M3 (Model 32) *Zero* fighter from the 204 *Kōkūtai* that was flown by PO2c Kenji Yanagiya on the historic 18 April 1943 mission to escort Admiral Yamamoto's aircraft. *Michael Claringbould*

J1N1-S (Type 11) *Gekkō* fighter, 210 *Kōkūtai*, Meiji Air Base in March 1945. This aircraft was flown by Lieutenant Minoru Toyonaga in defense of Nagoya City area. *Michael Claringbould*

J2M3 *Raiden* (Model 21), 302 *Kōkūtai*, Atsugi Air Base, the spring of 1945. This aircraft was flown by PO2c Mitsuyasu Yamakawa. *Michael Claringbould*

N1K2 *Shiden Kai* (Model 21), 343 *Kōkūtai*, Ōmura Air Base, July 1945. This aircraft was flown by Fighter 701 *Hikōtai* leader Lieutenant Takashi Oshibuchi. *Michael Claringbould*

SB2C-3 *Helldiver*, White 11, of VB-80, USS *Ticonderoga*, late 1944. It was flown by Lieutenant (junior grade) Charles Downey.
George Zielinski, A.G.P.

TBM-3 *Avenger*, White 2 of VT-10, USS *Intrepid*, spring 1945. It was flown by Lieutenant (junior grade) Grant Young.
George Zielinski, A.G.P.

Appendix I – Map of Japanese Navy Airfields and Seaplane Bases

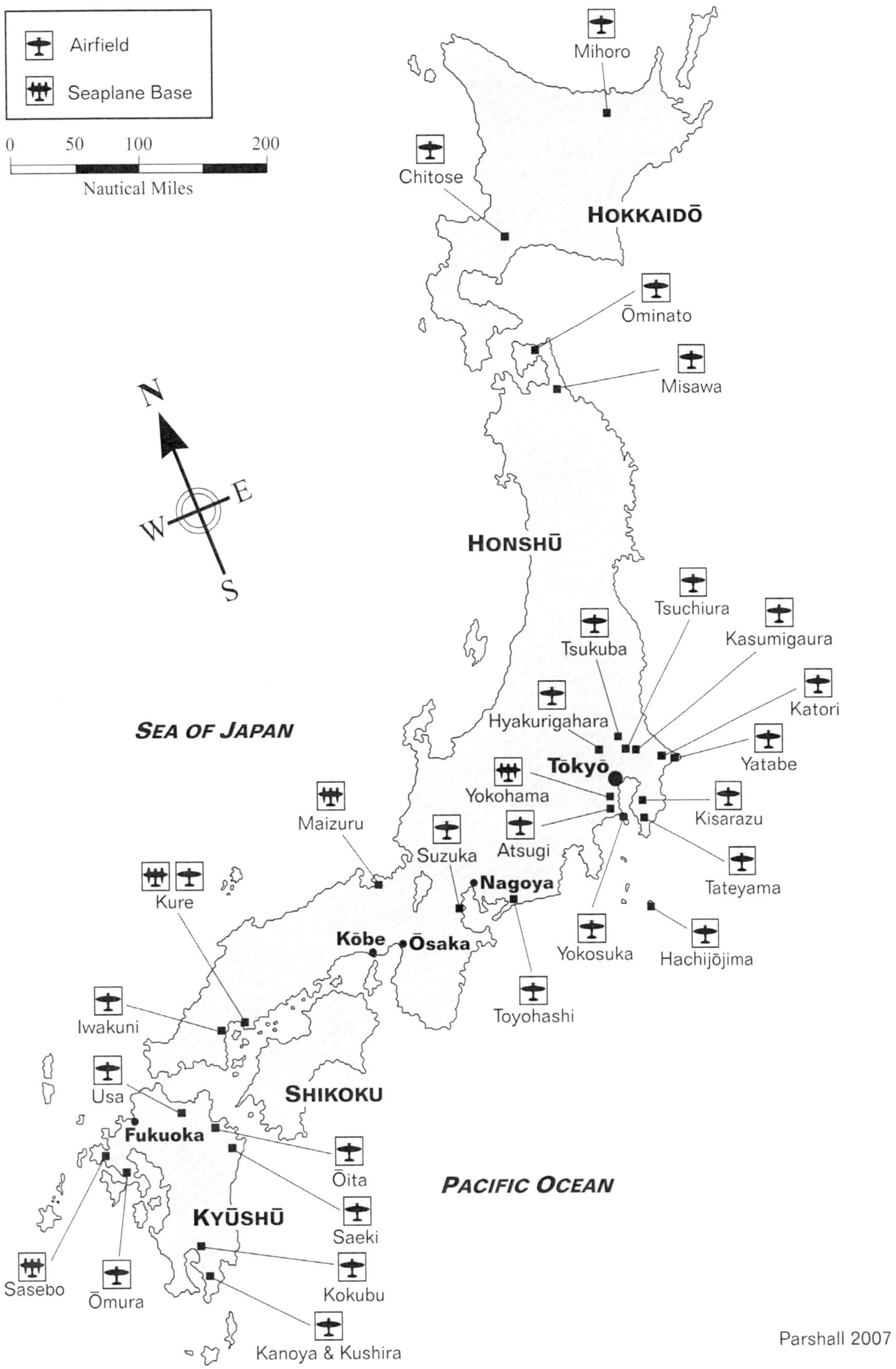

Appendix II – Glossary of Japanese Terminology

Buntai – A group or division; a *Hikō Buntai* – Flying group

Buntaichō – Division officer

Buntaishi – A leader of a section in a *buntai* but still under the command of the group's *buntaichō*

Chūkō – A casual unofficial term (see *rikkō*), being short for *Chūgata kōgekiki* (i.e. medium-size attack aircraft) It was used by Japanese Navy personnel as a casual-reference generic term to refer to twin-engine, medium-sized bombers such as the Mitsubishi Type 1 "Betty" to distinguish them from larger aircraft in the same category

Chūtai – A unit of six to nine aircraft

Chūtaichō – Commander of a *chūtai*

Hikōchō – Air Officer

Hikōkitai – Literally, it means "aircraft unit", but in vernacular English it signifies the flight echelon or aircraft echelon of the carrier (as opposed to other elements of the ship, such as the machinists in the boiler room, or the sailors manning the flight deck). This term was used for both carrier-based and land-based aircraft units

Hikōtai - Flight echelons of land-based units or carrier-based units

Hikōtaichō – Leader of a flight echelon

Hinomaru - The red disc that represented Japan which was painted on the aircraft's wings and fuselage. It also could be found on a number of other items such as Japanese flags, insignia, etc.

Hanchō – Leader of a small group

Kanbaku – The abbreviation used for a carrier bomber (dive-bomber), *Kanjō bakugekiki*

Kanbaku-tai – carrier bomber (dive-bomber) unit

Kankō – The abbreviation used for carrier (torpedo) attack plane, *Kanjō kōgekiki*

Kankō-tai – carrier attack plane unit

Kansen – The abbreviation used for carrier fighter, *Kanjō sentōki*

Kidō Butai – It literally means mobile force. The official designation for the Japanese carrier attack force

Kikusui – This is an ancient Japanese symbol dating to medieval times. Tradition ascribes it to Kusunoki Masashige, a warrior whose legendary loyalty to the emperor led him to die willingly in battle against overwhelming odds. It literally means "chrysanthemum and water" or "floating chrysanthemum", and was the overall name given to the series of *tokkō* sorties during the Okinawa campaign

Kōkūtai – A land-based Japanese Navy Air Group

Raigeki-tai – torpedo bomber unit

Rikkō – A more official term than *chūkō* used to describe land-based attack aircraft such as a Mitsubishi Type 1 "Betty" bomber. It is an abbreviation of *Rikujō kōgekki*, the proper designation for this category of aircraft

Saké – An alcoholic drink brewed mainly from rice

Shōchū – An alcoholic drink made from sweet potatoes

Shōtai – A unit of two to four planes (usually three)

Shōtaichō – Commander of a *shōtai*

Taichō – This term literally means "leader"

Tatami – Traditional Japanese flooring mats made of woven straw

Tokkō – a suicide attack mission using a variety of methods such as crashing into a target by airplane or a boat, which most westerners refer to as *kamikaze* but Japanese people normally use this term instead

Tokkō-tai – a Japanese military unit that carried out suicide attack missions.

Yokaren – Flight Reserve Enlisted Trainee Class System; *Otsu-yokaren* – Pilot trainees from the second grade of advanced primary school or middle school (fifteen years old or older); *Kō-yokaren* - Pilot trainees from the fourth grade of middle school (seventeen years old or older)

Appendix III – Japanese Naval Air Force Ranks[1]

Commissioned Ranks

Taishō	Admiral
Chūjō	Vice Admiral
Shōshō	Rear Admiral
Taisa	Captain
Chūsa	Commander
Shōsa	Lieutenant Commander
Tai-i	Lieutenant
Chū-i	Lieutenant (junior grade)
Shō-i	Ensign

Enlisted Ranks *(Before June 1941)*

Kōkū Heisōchō	Aviation Warrant Officer (WO)
Ittō Kōkū Heisō	Aviation Petty Officer, First Class (PO1c)
Nitō Kōkū Heisō	Aviation Petty Officer, Second Class (PO2c)
Santō Kōkū Heisō	Aviation Petty Officer, Third Class (PO3c)
Ittō Kōkūhei	Aviation Seaman, First Class (Sea1c)

Enlisted Ranks *(After June 1941)*

Hikō Heisōchō	Flight Warrant Officer (WO)
Jōtō Hikō Heisō	Flight Chief Petty Officer (CPO)
Ittō Hikō Heisō	Flight Petty Officer, First Class (PO1c)
Nitō Hikō Heisō	Flight Petty Officer, Second Class (PO2c)
Santō Hikō Heisō	Flight Petty Officer, Third Class (PO3c)
Hikō Heichō	Flight Leading Seaman (LdgSea)
Jōtō Hikōhei	Flight Superior Seaman (SupSea)
Ittō Hikōhei	Flight Seaman, First Class (Sea1c)

[1] Ikuhiko Hata and Yasuho Izawa, *Japanese Naval Aces and Fighter Units of WWII*, XIII.

Appendix IV – 302 Kōkūtai

The following three large group pictures are all related to the 302 *Ku* and show several of the veterans featured in the special section of the book. They include Ensign Sadaaki Akamatsu, Lieutenant Yukio Endo and Maintenance Lieutenant Minoru Yoshino. Yoshino-san was later promoted to the rank of maintenance lieutenant commander.

Page 266: Raiden *pilots of the 302* Ku *on 1 January 1945. Unfortunately, not every pilot is properly identified by name.*
Front row, left to right: Warrant Officer Futao Shikada, Ensign Sadaaki Akamatsu, Lieutenant (junior grade) Akai, Lieutenant Yoshinori Nakano (1st Fighter Group buntaichō*), Lieutenant Tominari Miyazaki (*hikotaichō, *2nd* Hikotai*) Lieutenant Commander Kushichiro Yamada (2nd Fighter Group* buntaichō*), Lieutenant Susumu Itoh, Lieutenant Sumio Teramura, Lieutenant Tsuboi, Lieutenant (junior grade) Ei Fukuda and Lieutenant Hiroshi Tsukada. Third row, far left (standing): PO2C Mitsuyasu Yamakawa. (Mitsuyasu Yamakawa Collection)*

Page 267: 302 Ku *mechanics, July 1944, in front the unit headquarters.*
Front row, left to right: Ensign Hirota (Raiden), *Lieutenant (junior grade) Tokugoro Osawa* (Raiden), *Lieutenant (junior grade) Shigeru Yamamoto* (Gekkō buntaichō), *Lieutenant Minoru Yoshino, Lieutenant Mitsuaki Masaki,* (Raiden, Zero *and* Zero *night-fighter* buntaichō), *Lieutenant (junior grade) Hideyoshi Suzuki* (Suisei), *Lieutenant (junior grade) Koji Hirose* (Gekkō) *and Lieutenant (junior grade) Yuji Konishi* (Zero *and* Zero *night-fighter). Middle row, left to right: Warrant Officer Adachi* (Gekkō), *unknown, Ensign Hiroshi Nakamura* (Raiden), *Ensign Shigeō Nakagawa* (Raiden), *Ensign Toshikazu Motobayashi* (Raiden), *Ensign Kiyotaka Sekiguchi* (Gekkō) *and Ensign Tadao Hatori* (Raiden). *Back row, left to right: First four men (unknown), fifth from the left, Warrant Officer Takezawa* (Zero *and* Zero *night-fighter), sixth and seventh men (unknown), Warrant Officer Shibasaki* (Raiden) *and Warrant Officer Hayashi* (Raiden). *(Minoru Yoshino Collection)*

Page 268: Gekkō *fighter pilots, aircrew and mechanics.*
Second row, left to right: Ensign Kiyotaka Sekiguchi (mechanic), Ensign Yukinobu Izumi, Ensign Kunio Nakayama, Ensign Toshio Kikuchi, Ensign Akira Isobe, Lieutenant (junior grade) Hiromasa Ōyama, Lieutenant Yukio Endo (Gekko buntaichō), *Lieutenant Shigeru Yamamoto* (Gekkō *mechanic* buntaichō), *Ensign Masayoshi Yokota, Ensign Tatsujiro Kawashima, Ensign Masao Shinohara, Lieutenant (junior grade) Mitsumasa Tomita, Lieutenant Sigeru Musha, Ensign Motonori Saibe and Ensign Mitsuo Kimura.*
Not pictured in this photo are Lieutenant Kiyoshi Kinoshita (hikotaichō, *2nd* Hikotai) *and members of the detachment that served at Hachijō Jima. (S. Bon Group)*

Select Bibliography

English-Language Books

Agawa, Hiroyuki. *The Reluctant Admiral: Yamamoto and the Imperial Navy*. Translated by John Bester. Tokyo: Kōdansha International, 1979.

Bradley, James with Ron Powers. *Flags of Our Fathers*. New York, NY: Bantam Books, 2000.

Brown, David. *The Seafire: The Spitfire that went to Sea*. Annapolis, MD: Naval Institute Press, 1973.

Cameron, R. Duncan. *Consolidated Stinson OY-1 BuNo 02747*. (self-published), 2012.

Dorr, Robert F. *B-29 Superfortress Units of World War 2, Osprey Combat Aircraft No. 33*. Oxford, UK: Osprey Publishing Ltd., 2002.

Francillon, Renē J. *Japanese Aircraft of the Pacific War*. Annapolis, MD: Naval Institute Press, 1970.

Glines, Carrol V. *Attack on Yamamoto*. Atglen, PA: Schiffer Publishing Ltd., 1993.

Hata, Ikuhiko and Yasuho Izawa. *Japanese Naval Aces and Fighter Units in World War II*. Translated by Don Cyril Gorham. Annapolis, MD: Naval Institute Press, 1989.

Lord, Walter. *Incredible Victory*. New York: NY: Harper & row, Publishers, Incorporated, 1967.

Moore, Stephen L. with William J. Shinneman and Robert Gruebel. *The Buzzard Brigade: Torpedo Squadron Ten at War*. Missoula, MT: Pictorial Histories Publishing Co., 1996.

Naito, Hatsuho. *Thunder Gods: The Kamikaze Pilots Tell Their Story*. Translated by Mayumi Ichikawa. Tokyo: Kōdansha International, 1989.

O'Donnell, Patrick K. *Into the Rising Sun*. New York, NY: The Free Press, 2003.

Pacific War Research Society, *Japan's Longest Day*. Tokyo: Kōdansha International, 1968.

Parshall, Jonathan and Anthony Tully. *Shattered Sword: The Untold Story of the Battle of Midway*. Washington, D.C.: Potomac Books, 2005.

Putney, Captain William W. Putney D.V.M., USMC (Ret.). *Always Faithful: A Memoir of the Marine War Dogs of WW II*. Washington D.C.: Brassey's Inc., 2003.

Ross, Bill D. Ross. *Iwo Jima: Legacy of Valor*. New York, NY: Vanguard Press Inc., 1985.

Sakaida, Henry. *Imperial Japanese Navy Aces 1937-45, Osprey Aircraft of the Aces, No. 22*. Oxford, UK: Osprey Publishing Ltd., 1998.

Sakaida, Henry and Koji Takaki. *Genda's Blade: Japan's Squadron of Aces, 343 Kōkūtai*. Surrey, UK: Ian Allan Publishing Ltd., 2003.

Tillman, Barrett. *Helldiver Units of World War 2, Osprey Combat Aircraft No.3*. Oxford, UK: Osprey Publishing Ltd., 1997.

Umeda, Atsushi. *Japan: A Bilingual Atlas*. Tokyo: Kōdansha International, 1991.

Werneth, Ron. *Beyond Pearl Harbor: The Untold Stories of Japan's Naval Airmen*. Atglen, PA: Schiffer Publishing Ltd., 2008.

Wolf, William. *13th Fighter Command in World War II: Air Combat over Guadalcanal and the Solomons*. Atglen, PA: Schiffer Publishing Ltd., 2004.

Japanese-Language Books

Bungei Shijyu (compiled by). *Shōgen Ōka Tokkōtai Ningen Bakudan to Yobarete*. Tokyo: Bungei Shijyu, 2005.

Hata, Ikuhikō. *Hachigatsu Jugonichi No Sora*. Tokyo: Bungei Shijyu, 1978.

Kaigun Jinrai Butai Veterans' Group (compiled by). *Kaigun Jinrai Butai*. (1996)

Kōdachi, Naoki. *Reisen Saigono Shōgen, Vol. 1.*Tokyo: Kōjinsha, 1999.

Kōdachi, Naoki. *Reisen Saigono Shōgen, Vol. 2*. Tokyo: Kōjinsha, 2000.

Sagara, Shunsuke. *Ah, Atsugi Kōkutai - Patriotic death of a Samurai*. Tokyo: Kōjinsha, 1993.

Watanabe, Yoji. *Shashin Shi 302 Ku*. Tokyo: Kōku Fan, 1997

Index

Aircaft Types

Aviation Squadrons and Groups

The Author

Ron Werneth has been interviewing American military veterans and recording their stories since he was a teenager. For more than twenty-five years, he has made it his personal mission to record and preserve the stories of World War II veterans from all sides of the conflict. Embarking on a two decade journey that included living in Japan for several years, he succeeded in meeting and interviewing the last surviving Japanese WWII Navy pilots and aircraft mechanics. For many, this was their the first opportunity to share their stories – especially with English-language readers.

In 2008, Ron Werneth published *Beyond Pearl Harbor: The Untold Stories of Japan's Naval Airmen,* considered to be a groundbreaking book on World War II aviation. Ron Werneth has contributed regularly to numerous aviation and historical publications in the United States, Europe and around the world.

Ron Werneth lives in Chicago with his wife and daughter.

www.ingramcontent.com/pod-product-compliance
Lightning Source LLC
LaVergne TN
LVRC101546220926
848486LV00041B/387

9781943492442